The Brilliance of Birds

The Brilliance of Birds

A NEW ZEALAND BIRDVENTURE

SKYE WISHART & EDIN WHITEHEAD

PENGUIN BOOKS

Contents

COVER **Kea.**
PREVIOUS PAGES Page 1: kea;
pages 2–3: white-fronted terns;
pages 4–5: Australasian gannet;
page 6: white-fronted tern.

Introduction

— *Skye Wishart*

For 80 million years, New Zealand birds never saw a single mammalian tooth or claw. Their only predators were fearsome birds of prey, hunting from the air by sight and not by smell; New Zealand was largely safe on the ground. Many birds could now abandon flying to save energy or grow larger. Some converted to ground nesting, or came out to feed only at night, and some became statue-like when disturbed. In this land of feathers, birds could occupy niches that are held by mammals in other ecosystems; for example, the role of insect-eating leaf-litter forager was taken by the kiwi, instead of the hedgehog or anteater. Primitive species like New Zealand wrens, tuatara and ancient frogs could survive while their counterparts in other countries became extinct with the evolution of predators. New Zealand also became the seabird capital of the world — more than 80 species breed here, and more than a third of them are endemic.

New Zealand birds, in their long isolation from the rest of the world, also acquired unique and often bizarre characteristics: the wrybill is the only bird in the world with a beak that curves sideways, always to the right; the kiwi, with its shaggy, mammal-like feathers, has nostrils at the end of its impossibly long beak; the owl-like kākāpō is the world's heaviest parrot and the only nocturnal, flightless one, too; the audacious kea is the world's only alpine parrot; and the hihi can mate face to face. Before humans arrived in this unique ecosystem, there were about 245 bird species breeding in New Zealand and 71 per cent of them were endemic.

Some birds have been resident here since Zealandia split from Gondwana — birds such as the New Zealand wrens, and perhaps the wattlebirds. Other birds made their way to New Zealand on the wing, mostly from Australia, but then evolved with a New Zealand twist: the stocky South Island takahē was once a pūkeko-like bird; the black stilt's ancestor was black and white. Even the extinct Haast's eagle, with its 3 m wingspan, almost 18 kg bulk and taste for moa, was recently determined to have had similar DNA to the 1 kg Australian little eagle (that is, they share the same common ancestor, but the Haast's eagle evolved to be larger because of a lack of competition and an unexploited moa population).

Then humans arrived. They destroyed bird habitats by burning and clearing forests, and draining wetlands too. Māori hunted moa, geese, adzebills and others to extinction and brought with them kiore (Polynesian rats), which ate the eggs and chicks of ground nesters. Europeans cleared even more land, helped hunt more birds to extinction and brought bigger rats with them on their ships. The colonists began to import species, believing that New Zealand was a clean slate and desperately needed the species familiar in Europe: game birds, song birds, insect eaters and mammals. They introduced rabbits for hunting, and then ferrets, weasels and stoats to control the exploding rabbit population. These introduced animals promptly started killing New Zealand's defenceless and flightless birds instead. Mustelids, possums, cats, rats and deer are still today decimating the remaining bird populations and their habitats. At least 51 birds (41 per cent of all endemic species) have become extinct since human settlement, and many more are on their way.

But modern conservation efforts are slowly paying off for many birds. New Zealand has become a world leader in making offshore islands predator-free, as places to grow populations of threatened birds: nearly half of the world's native species that have benefited from the eradication of invasive mammals from offshore islands are New Zealand species (notable examples are kōkako, takahē and saddleback, which have come back from the brink of extinction). Mainland sanctuaries, such as Zealandia, have also created havens for native birds that create an overflow effect, populating surrounding areas where there is pest control. These places have been so good at boosting populations that the birds are outgrowing them — some takahē have now been released into the (intensively pest-controlled) wild in Kahurangi National Park, near Nelson.

While the average New Zealander might never have the privilege of seeing a kākāpō or Chatham Island snipe, regular birds are the ever-present backdrop of our lives: swooping on coastal bluffs, dashing through bush, scavenging for scraps on the streets and primping in our backyards — occurrences so common we often pay no attention. But to learn more about all kinds of birds is to discover that even New Zealand's most everyday avians have some of the weirdest and most wonderful stories life has to offer: drama, adventure, fierce loyalty and highly peculiar habits. To spot pūkeko and know that they're raising their chicks as a tribe, and are so ferocious they can take down a harrier. To listen to a territorial tūī switching its whirring sound on and off by angling special feathers. To see a tubby godwit and know it has shrunk its digestive organs impossibly small so it can fly nonstop to Alaska.

To know more about birds is to feel something for them: to notice even the everyday ones and enjoy observing them going about their busy lives, but also to be inspired to help protect our more precious birds from their slide towards extinction. The absence of these creatures, perfectly evolved in their weird and wonderful ways to occupy various roles in the ecosystem, would make New Zealand less charming, captivating and complete, and certainly less melodious.

The te reo Māori names in this book are those most commonly used and available to the public. These were drawn from three trusted New Zealand websites: New Zealand Birds Online, Māori Dictionary and Te Ara — The Encyclopedia of New Zealand.

The New Zealand Status defines where the species are found in the world. Endemic species are found in New Zealand and nowhere else; native species are found in New Zealand and other countries; and introduced species have been brought to New Zealand by humans from other countries.

The Conservation Status refers to the New Zealand Department of Conservation threat classification for New Zealand species. 'Threatened' species fall into Nationally Critical, Endangered or Vulnerable categories — those that are at the greatest risk of extinction. 'At risk' species are at a lower level of threat and can be Declining, Recovering, Relict or Naturally Uncommon. Species that are not considered at risk of decline or extinction fall into the categories of Not Threatened or Naturalised (for introduced species).

Australasian crested grebe

Frilly-headed high-country lake diver

Māori names	Latin name	New Zealand status	Conservation status
Pūteketeke, kāmana	*Podiceps cristatus*	Native	Nationally vulnerable

New Zealand's largest species of grebe specialises in lake diving while looking good. Year round, both male and female maintain their glorious chestnut facial ruff and spiky black crest while they plunge into chilly high-country lakes. Reappearing with an impossibly large fish in their beak, they'll deftly gulp it down a slender and elegant neck. The handsome crest is known to Māori as tikitiki — a word also used for the topknot worn by a high-ranking man.

Crested grebes are almost always in the water because their legs, with their funny lobed feet, are set so far back on the body that they are useless on land: the scientific name *Podiceps* means 'bum-legs' or 'arse-legs'. So they are phenomenal swimmers — their efficiency is incredible — but on the extremely rare occasions they walk, according to zoologist and grebe specialist John Darby, they move like a 90-year-old man with a walking stick. Although they have small wings, they sometimes fly hundreds of kilometres from iced-over high-country tarns down to lowland lakes

These birds are almost always in the water, because their legs are useless on land.

for the winter, but this is hardly ever seen, probably taking place at night.

Their year-round costume (unique within grebes) comes in handy during mating season, for crested grebes know how to put on a good water show. They rise up in the water, chest to chest, growling and cooing, and turning their heads side to side in unison to show off all angles of their fluffed-up crests and cheek ruffs. They also demonstrate what incredible nest partners they'll be by offering nesting material in their beaks — even if it is just a stick dripping with lake slime.

There's another use for the plumage. Like other grebes, this grebe will eat its own feathers from its breast and flank, and adults will even offer them to their chicks, after dipping them in water. It's uncertain why they do this — reasons range from protection against parasites to acting as a buffer against sharp fish bones (the bones are regurgitated along with the feathers as a pellet).

Also like other grebes, this grebe ferries its chicks around on its back. Even though the chicks can swim, you'll often see their little black-and-white striped heads poking up between a parent's folded wings. Each parent can carry up to three chicks. This may keep the young warm and also protect them from predators, such as eels. Occasionally, adults will dive with the chick still on their back, but usually they'll shake them off into the nest first — an action so ingrained that the parents begin doing it when heading off for a dive even before their chicks have hatched.

Crested grebes are found in Australia, too, but in New Zealand they are found only in the South Island in lowland lakes and subalpine tarns, mainly in the Otago and Canterbury regions. They prefer large, clear glacial lakes, where they forage for fish and sometimes invertebrates. Fossil records and the remains of Māori middens show they were once widespread in the

Small but powerful wings allow Australasian crested grebes to cover large distances, but are also useful in territorial disputes.

North Island; today, they're threatened, with perhaps fewer than 1000 birds in New Zealand (although, since they're always in the water, it's impossible to count them using leg bands).

Crested grebes often use vegetation at the water's edge — such as rushes, sedges or willow — to anchor their messy floating nests, which they top up with waterweed every time they relieve their companion on the nest. They start incubating the moment they lay their first egg, as opposed to waiting until more are laid. Since the eggs are laid two days apart, some chicks hatch a lot later than others. This is bad news if the parent is scared away before incubation is complete, as it will abandon any unhatched eggs.

One of the biggest factors affecting grebe numbers — apart from predation by mustelids and cats — is thought to be simply having a place to make a nest. If they're on a lake with a hydro-electric scheme or motorboats generating lots of wake, nests may be swamped or left high and dry on the banks, both a disaster for the eggs. John Darby has enabled a small population of grebes to nest and successfully fledge more than 200 chicks on Lake Wanaka by making them a nesting platform of pool noodles, plywood, weed matting, branches and lakeweed, anchored to the marina. While grebes are normally very shy and can abandon their nests if approached, this population seems to tolerate the busy area.

When chicks
are small,
parents take
turns carrying
them and
feeding them
delicate
morsels of
larval fish,
Lake Wanaka.

Australasian gannet

Daredevil diver with magic eyes

Māori names	Latin name	New Zealand status	Conservation status
Tākapu, tākupu	*Morus serrator*	Native	Not threatened

From up to 20 metres in the air, gannets plunge into the water at insane speeds to catch small fish and squid — but being a daredevil can cause accidents. Occasionally, these gorgeous birds die from neck and head injuries when they collide with other gannets in the water, especially when two birds target the same fish with their lethal wedge-shaped bill.

The life of an Australasian gannet is dramatic from the get-go. As soon as they can fly, the mottled young birds at their New Zealand colonies head for Australia, making the almost 3000 km migration direct. They stay there for their first three or so years before returning to their breeding colony — the bird version of a young person's OE.

By the time they return, both males and females have blossomed into adult plumage: mainly pure white, with metallic blue eye shadow and a yellow-washed head. From July they make their nests in jam-packed, smelly, squawking colonies on headlands and islands around mainland New Zealand. There are 26 colonies on the east coast and three on the west, the largest and most accessible including Cape Kidnappers, Muriwai and Farewell Spit.

The life of an Australasian gannet is dramatic from the get-go.

If nesting on a new patch of ground, gannets will collect grasses and seaweed into a mound, which over time becomes a hardened white mass due to all the guano. After years of breeding seasons, these nests form a dense network of raised mounds, each representing a tiny territory that the male repossesses every year and fiercely defends, waiting for his female to arrive. A breeding pair can heavily wound any bird, human or other animal that comes too close to their single egg or chick. (If you see two eggs, it's likely two females laid in the same nest.) The female and male take turns on the nest, using their webbed feet to keep the precious contents warm. Each adult flies in from foraging with a loud call, somehow recognising its mate among the hundreds of others at the colony, who all look exactly the same to the human eye. Once it has landed, the two birds perform a bill fencing ceremony to help

reinforce the pair bond. The subsequent task of delivering the food to the chick is not glamorous: adults don't completely regurgitate their food, so the chick has to push its bill right into the parent's throat to fish it out.

Gannets fish right in front of their colonies, or fly up to 300 km for a good food source. Favourite foods include garfish, squid, mackerel, anchovy, kahawai and mullet. They have amazing aerial vision: not only can they spot a fish through the shimmering ocean surface, but they also magically switch to aquatic vision when they hit the water. While we humans have to put on goggles to create a pocket of air between our cornea and the water if we are to see anything clearly, gannets have another solution. Research by Gabriel Machovsky-Capuska found gannets change the shape of the lens of their eye from oval (far-sighted) to

Gannets are endlessly gentle with their ferocious beaks when greeting their partners or preening their chicks. The beak is serrated along the inside edge for gripping prey.

spherical (near-sighted) within a tenth of a second after submerging. With a more spherical lens, light bends more, keeping a sharp image focused on the retina. This lets them see well enough underwater to sometimes catch four or five fish as they swim around. It's the wild west down there, too, with some gannets stealing other's catches. The spherical lens is common in marine animals like seals and penguins, but it's the gannet's speed of change that is impressive. It's been said that gannets get cataracts when they're older, as a result of diving so much, but this is just a rumour — these birds can handle their diving into a ripe old age.

Gannet pairs are often held up as a romantic example of lifelong love, but this is another rumour. A 2009 study by Stefanie Ismar found that in some pairs, the male was not the father of the chick — it may instead have been a roaming male. Ismar found that around 40 per cent of pairs had switched partners between two seasons. However, those that had stayed together fledged more chicks, perhaps because they had learned to cooperate. It's thought that gannets switch partners if one is too late getting back to the colony — better to re-pair than risk not breeding at all. It could also be a way of minimising inbreeding.

Australian coot

Fierce parent with outlandish feet

Māori name	Latin name	New Zealand status	Conservation status
n/a	*Fulica atra australis*	Native	Naturally uncommon

The term 'old coot', slang for a doddery old man, aptly fits the goofy way this bird rocks its head back and forth when swimming. And the saying 'Bald as a coot' could easily refer to the newly hatched chicks, whose bald pink heads shine grotesquely through striking 'fire-tipped' feathers.

The Australian coot looks a bit like a pūkeko, with the shield on its forehead — white, versus the pūkeko's red — in gleaming contrast to its dark grey body. But it has completely bizarre feet. Instead of the pūkeko's long slim toes, Australian coots have flattened, scalloped lobes flanking their short grey toes. Although these fleshy appendages look ungainly on solid ground, they're just the right kit for aquatic life. When the swimming coot brings its foot forwards underwater, the lobes fold and streamline, and with a backwards stroke they unfold to create resistance to the water and give the bird the powerful propulsion it needs to swim and dive. This 'webbing' also allows the bird to patter across the water as it takes off — coots prefer not to fly, but they are strong fliers when they do, neck stretched out and feet trailing behind. The legs look strange on land, as though they are set too far back on the body, but this doesn't hinder the coot's walking, and does add to its swimming prowess.

New Zealand originally had its own, much larger coot species, which became extinct soon after human contact.

As for the Australian coot, it was first found (or rather, shot) in New Zealand in 1875, but it wasn't until 1958, when it was sighted breeding on Lake Hayes in Otago, that the species was considered to be a self-introduced native. Today the Australian coot enjoys full protection in New Zealand, with some 2000 birds scattered nationwide (except Northland), wherever there are lakes and ponds filled with waterweeds and edged with reeds.

The Australian coot is a subspecies of the Eurasian coot of Europe, Asia and Africa. But while the Eurasian birds eat insects, molluscs, fish and amphibians, their Australian cousin is mainly vegetarian, eating algae, shoots, seeds from the surface, and the tips of waterweed fronds deep underwater. (It will, however, sometimes eat the eggs of other birds.)

Allopreening, where birds preen each other's feathers, reinforces social bonds.

Don't be fooled into thinking this vegetarian is a pacifist, however. It becomes a territorial beast in the breeding season.

To dive, the coot cocks its head sideways to sight the waterweeds with one red eye, then up-ends its bum and disappears straight down.

Don't be fooled into thinking this vegetarian is a pacifist, however. It becomes a territorial beast in the breeding season, when adults pair up and the female lays her eggs in a floating nest of twigs anchored to branches or reeds. In Europe, coots with nestlings can be so aggressive that they take on larger birds — and win. One observer in the Netherlands saw a mother coot drowning a bird of prey that had swooped down hoping for a snack — she pecked and wing-flapped and used her body weight to push the buzzard under the water and waterlog its feathers until it drowned. In New Zealand, Australian coots are known to rush at ducks, and even swans, if they swim too near the nest.

Both parents incubate, and they get help from their older chicks in raising the younger ones. The babies have jet-black down except for the flame-hued head plumage, which fades in the first few days after hatching. While this has not been studied in Australian coots, in American coots the females favour the chicks with the stronger head colouration — so even though the more brightly coloured chicks might draw attention from predators, they get more food and parental care, so are more likely to survive.

Coot nests are generally secretive affairs, but those tucked beneath the willows at Western Springs, Auckland, are easily visible and the birds don't seem to mind watchers from the shoreline.

'Fire-topped' coot
chick, Western Springs,
Auckland.

Australian magpie

Quardle-oodle-ardling dive bomber with an overblown reputation

Māori name	Latin name	New Zealand status	Conservation status
Makipae	*Gymnorhina tibicen*	Introduced	Naturalised

This large, strutting bird with a stark and glossy black-and-white suit is hard to confuse with any other bird — especially when it's raucously chortling at dawn and dusk or dive-bombing anyone who wanders past its nest in the breeding season.

The Australian magpie was introduced to New Zealand from across the ditch in the 1860s-70s to control insect pests in pasture — it has an effective strategy of walking (rather than hopping) along the ground, listening and watching for invertebrates just under the grass, and even perhaps sensing their vibrations. Magpies were also introduced to Fiji to control stick insects on coconut palms but failed in their quest; stick insects are up in the trees, and magpies feed only on the ground. However, magpies sure did their job for New Zealand farmers, gulping down all manner of prey — worms, caterpillars, wasps, spiders, snails and

more — before hoicking up the indigestible parts in compact little pellets. They'll also eat seeds, carrion, mice, lizards and sometimes other birds. But in time they showed another side of their nature: complaints flooded in about them driving away native birds and attacking people, and their protection was removed in 1953.

The magpie certainly has a hostile reputation when it comes to defending its family — it's famous for swooping — but in reality only one in ten male magpies will swoop down on humans. When they do, it's terrifying. They'll scold and scold, and if that doesn't deter you, they'll start the aerial attacks — repeated dive-bombs

with a sharp snapping beak aimed at your head, particularly your eyes. Magpies are extremely smart, with incredible memories — research has shown they can recognise a human face (if you wear different clothing, they still know who you are). Considering magpies can live in one territory year-round for up to 20 years, they can get to know their regular residents pretty well — and target them during breeding season if they see someone they particularly dislike within 50-100 m of their nest with their pink, naked and blind chicks yet to fledge. Australian magpies can hold a grudge,

too — for example, if a chick falls out of the nest and the adults see a human trying to put it back in, they'll interpret that as a predatory action, and that human will forever be marked as the enemy. In their native Australia, where magpies are much more present in urban centres than they are in New Zealand, cyclists have been seen positioning spiky cable ties on their helmets, wearing sunglasses on the backs of their heads, carrying open umbrellas, and walking their bikes in the territories of particularly aggro magpies.

While a bonded male and female will

Young magpies have mottled grey plumage before they moult into their stark black-and-white adult feathers.

hold their territory for what seems like forever, if one partner dies, it will be replaced almost immediately. Even if the female has chicks, her new male will jump right on in and help feed them as if they were his own.

Some magpies, however, don't have a territory to defend. These birds hang out in nomadic flocks of up to 80-100 birds, both males and females. This is a motley crew of juveniles (with a second plumage of mottled grey-and-brown feathers along with the black-and-whites) who've been evicted from their birth territories, along with adults who haven't yet paired up or pairs that haven't been able to find a territory.

Other flocks will be very territorial if they're part of a cooperative breeding group. More common in Australia, these groups have a main female and male pair who do the breeding, nesting and chick-raising; but the larger the group, the more often this pair will have cheeky relations on the side with some of the members — and so it's not entirely certain who the parents are.

The magpie has a complex variety of calls. The main one, a warble containing up to 893 syllables, can be performed for hours. The bird will even add in mimicry — from other bird species to dogs, sirens, cars, you name it. It also has a suite of alarm calls, which can be short and harsh or very complex, depending on what it is alarmed about. The warble in particular is immortalised in New Zealand poet Denis Glover's famous 1941 poem 'The Magpies' about the Depression of the 1930s, when many, including farmers, were doing it tough:

But all the beautiful crops soon went
to the mortgage man instead
and *Quardle oodle ardle wardle doodle*
The magpies said.

The magpie's scientific name *tibicen*, appropriately, is Latin for 'flute player'. Flautist it may be, but the bird received its common name because of how much it looked like the European magpie. In reality they're in different families. In Australia there are eight subspecies, but the magpies in New Zealand come in just three: most of the population belong to two white-backed forms (one with the name of *tyrannica*), but there is also a black-backed subspecies that is particularly dominant in Hawke's Bay and Canterbury. These

Magpies in New Zealand come in white-backed and black-backed varieties (and in all manner of intermediates).

subspecies were imported from different parts of Australia, Tasmania and southern New Guinea — but they all crossbreed so are essentially just one species.

The Australian magpie is party to a few myths. One is that it is a major pest to native bird populations, terrorising tūī and kererū among others, killing smaller birds and stealing chicks and eggs from the nests. Research in the early 2000s by John Innes, Dai Morgan and others on thousands of hectares of land across New Zealand failed to provide evidence for this. They found our birds do tend to avoid magpies, and lie low so as not to be seen — and in fact one of the studies showed that magpies attacked at least 45 species, including native birds — but when the researchers removed magpies from large areas of land, they couldn't find any change in the overall numbers of species and abundance of birds that were present there. Their conclusion: magpies are attacking such a tiny percentage of the birds they encounter that it doesn't make much difference to the overall bird population. One possible reason for the perception of their impact is that they are more likely than rats, cats and mustelids to be caught in the act: they're loud,

irritating and eye-catching, and active during daylight hours — even if they don't devastate wildlife like those nocturnal predators do. In some cases, their attacks could be of benefit to native birds — if they see a raptor such as the swamp harrier, magpies will sound the alarm and attack the bird in a pair or a mob, driving it out of their territory and inadvertently protecting other native birds in the vicinity.

Another myth is that of kleptomania. Magpies have for centuries had a reputation for being attracted to, and stealing, shiny things. But this pertained to the European magpie, and in any case it has been debunked as an urban myth: a 2014 UK study found that the birds were ambivalent towards, or even afraid of, bright shiny things.

There's also an unlikely side to the magpie's maniacal and bullying nature: play. Juveniles and adults will roll around on their backs, playing with objects and chasing each other for leaves and feathers: something to which many people who've kept pet magpies would attest.

Bellbird

Silver-tongued nectar sipper

Māori names	Latin name	New Zealand status	Conservation status
Korimako, kopara, makomako	*Anthornis melanura*	Endemic	Not threatened

Bellbirds are the ultimate crooners, and their song is arguably one of the most beautiful sounds to come out of New Zealand's forests. Their bell-like notes are thought to be those famously described by English naturalist Joseph Banks in Queen Charlotte Sound when James Cook first sailed to New Zealand.

Banks wrote, 'I was awakened by the singing of the birds ashore . . . They seemed to strain their throats with emulation, and made, perhaps, the most melodious wild music I have ever heard, almost imitating small bells, but with the most tunable silver imaginable . . .' Similarly, a 1922 article by J. Cowan in the *Auckland Star* waxed lyrical about the morning chorus in Te Urewera: '. . . minute after minute went by and the ringing, chiming and scale running grew louder.

It grew into a magic ecstasy of song, with a fairy touch about it . . .'

This effusive dawn chorus is mainly the domain of the male bellbird, but the female can sing her heart out, too — usually to advertise her territory to other females. Her song is different to the male's, according to work by Dianne Brunton's team at Massey University. When bellbirds are tiny chicks, males and females have the same 'baby babble', but as they get older, they learn gender-appropriate

In spring, bellbirds will appear with pollen-coated foreheads, colour-coded by the plant species they've been sipping nectar from — blue for tree fuchsia, orange for flax.

The bellbird or korimako was special to early Māori: it was a spirit bird, a messenger between humans and gods.

songs that they hear from adult birds, and practise them until perfect (young males might sing female syllables at first, and vice versa, but social reinforcement quickly weeds this out). The females have slightly fewer syllables, and tend to sing just one song and then stop. The males, on the other hand, will sing endlessly on, one song after another.

There are regional variations, too: the dialect from a population on one island in the Hauraki Gulf, for example, will differ from another nearby. These birds can, nonetheless, still interbreed (unlike kōkako from different regions, which can't understand each other's dialects). The tūī is another bush bird whose voice is confusingly similar, but it can be identified by the odd harsh click or grunt breaking up the song, while the bellbird keeps it smooth.

Bellbirds are flitting, restless birds and are found in forest and scrub in most parts of the country. In the 1860s they almost completely disappeared north of Waikato, apart from the Coromandel and offshore islands; it's not known why, but disease and introduced mammalian predators are contenders, and mammalian predators still keep the bird numbers low. Now, some predator-free areas are regaining their bellbirds. In the Auckland region, for instance, Tawharanui Open Sanctuary, Motutapu and Shakespear Regional Park are being colonised again by birds translocated from predator-free islands like Little Barrier/Hauturu and Tiritiri Matangi. Bellbirds are common in

The bellbirds on the Poor Knights Islands are a unique subspecies with a striking local song dialect.

the South Island and on Stewart Island/ Rakiura and the Auckland Islands. They live primarily in native forests but will venture into human environments: in plantations and shelterbelts; along riverbanks; in gardens, parks and orchards.

Bellbirds are nectar fiends. Every day, they travel for kilometres, busily working trees from bottom to top, spiralling up the trunks and branches, using their brush-tipped tongues to sip the nectar from deep inside flowers, trying to avoid the tūī, which aggressively dominates nectar sources. Bellbird favourites include kōwhai, flax and the exotic flowering gum. After sipping on a flower they may have a dusting of pollen painted on their head and chest, which they'll transfer to the next flower they visit. Bellbirds also glide around catching insects and spiders from trees — especially during the breeding season, for extra protein — and eat small fruits (they're a great spreader of seeds) and honeydew off beech trees, although introduced wasps steal this delicious food source.

A pair defends the same territory every year while breeding (but may sneak out to take advantage of a tree full of nectar).

Female bellbirds have a smart white moustache strip and are a more muted olive green than males.

Out of the breeding season, bellbirds are usually nomadic and operate alone. Males are larger than females and will often hog the nectar sources. Bellbirds fly with a quick whirr — especially the male, which, like the tūī, can turn on a louder whirr when defending his territory, by changing the angle of his notched wing feathers.

The bellbird or korimako was special to early Māori: it was a spirit bird, a messenger between humans and gods, and was often offered as a sacrifice. When a high-ranking baby boy was to be named, a bellbird was released. In some areas the tohunga (priest or learned person) would eat a cooked bellbird, in the belief that the baby would become a great orator, just like the bellbird. If someone was a graceful orator, or an incredible singer, Māori would say 'he rite ki te kopara e ko i te ata' — 'it is like the bellbird singing at dawn' (kopara being one of many Māori names for the bird). To catch bellbirds, Māori would spear them or snare them in a trap that was sometimes baited with nectar-rich flowers, such as pōhutukawa. They'd also lure them by imitating their song, blowing on a leaf held between the lips.

Black swan

Vocal Australian that forms civil unions

Māori names	Latin name	New Zealand status	Conservation status
Kakīānau, wāna, wani	*Cygnus atratus*	Native	Not threatened

Unlike the classic white mute swan from Europe, this black bird is a big mouth — and is found bugling, crooning and whistling on large lakes, lagoons and estuaries up and down the country.

These loud black swans have modern families — while most will pair up with the opposite sex, male-on-male relationships are very common. To have chicks, a male pair either steals someone's nest, or forms a threesome with a female until she lays the eggs, after which the paired birds drive her away and bring up the cygnets themselves (although sometimes this *ménage à trois* will last). Two dads work well in swans, with fantastic territory defence creating a safe home for the chicks. Swans will beat away intruders with their wings, which can bruise and occasionally even break bones in small children and elderly people.

And that is how swans fight each other, too: a duel with the wrist bones. Before resorting to a scuffle, however, two male swans will perform a 'parade' display on a territorial border, though it may look to us like a romantic union (males and females look alike apart from a very slight size difference). The fierce rivals push against each other while swimming up and down an invisible line, with wings slightly raised and ruffled, white feathers showing, necks curved. After about 10 minutes, this either escalates into a fight or they'll go back to their partner and show off in a 'triumph ceremony'.

There used to be more black swans in New Zealand — about 100,000, with 70,000 on Lake Ellesmere alone — until the *Wahine* storm of 1968 dampened the population to less than 10,000 by annihilating the beds of aquatic plants.

Swans can be fierce in
defence of their cygnets.

Although originally
Australian, the black swan
is now considered a native
species in New Zealand.

These loud black swans have modern families . . . male-on-male relationships are very common.

In other areas, where lake pollution affects aquatic plants, swan numbers have also dropped; nationally, there are now around 50,000. Before this, they were regularly culled because of their voracious appetite for pasture and their copious droppings. They are still controlled in areas where they eat too much pasture or foul the water enough for it to become a health hazard to swimmers (the Rotorua lakes, for example, have huge numbers of swans fouling picnic and swimming spots). Only partially protected, black swans may be hunted in duck-shooting season, with a licence. Some say they taste a little like roast beef, but they've never been as popular in New Zealand as duck. On the Chatham Islands, swan eggs are harvested from the vast Te Whanga Lagoon and by all accounts make a good sponge cake.

Black swans can live 30-40 years and nest in territories or colonies, depending on how much food is around. They can lay more than one clutch of eggs in a season. On a huge shallow lake, with lots of waterweed and surrounding pasture, they form busy colonies, nesting not too far from the water on a large mound made of grass, weeds and leaves, where they defend as far as they can peck while sitting. But on a deep lake with scant food resources, a pair will fiercely defend a larger territory and closely guard their five or six cygnets. In colonies, relationships can be a bit loose: a female may dump her eggs in another nest, and her chicks may also not belong to her partner if there is an opportunistic male nearby. For the first couple of weeks the cygnets in colonies might gang together in groups of up to 40, with a couple of guard adults. The cygnets often ride on a parent's back when they are very small, their fluffy grey heads poking out from between the adult's wings.

New Zealand's black swans were introduced from Melbourne from the 1860s onwards. They came to both main islands, including one lot released onto Canterbury's Avon River (to get rid of the watercress so that the current would flow faster). But the population grew so rapidly and over such a wide area, that it's thought, coincidentally, at the same time some flew themselves over from Australia. Black swans migrate within New Zealand, moving between nesting areas and moulting areas (at the moulting areas, they are flightless for about a month). When they do fly, swans need a decent amount of open water to run across the surface with wings slapping the water.

The black swan is smaller than the European white or mute swan, which also was introduced to New Zealand during the nineteenth century. To early Europeans, the existence of black swans was once thought to be impossible. In the first century CE, the Roman poet Juvenal wrote of a *rara avis in terris nigroque simillima cygno* — 'a rare bird in the lands and very much like a black swan' — and this phrase was used in sixteenth-century London to describe something impossible. But when seventeenth-century Dutch

explorers to Western Australia became the first Europeans to see black swans, these birds came instead to represent that which was believed impossible but might someday be disproven. 'Black swan theory' describes something that happened as a surprise, and with hindsight is later wrongly rationalised as though it had been expected. Australian black swans were themselves introduced to Western Europe in the late eighteenth century as ornamental birds, but escapees bred in the wild and there are now feral populations in England, Belgium, the Netherlands, France and Italy.

New Zealand once had its own black swan: the poūwa (*Cygnus sumnerensis*). It existed on the mainland and the Chatham Islands when Polynesians first arrived, but was soon hunted to extinction. DNA and bone research by Nic Rawlence in 2017 suggested these swans arrived from Australia one or two million years ago, evolved to have longer legs and smaller wings, and were chunkier at around 10 kg (compared with 6 kg for Australian swans). It's a textbook example of evolution in island ecosystems that are free of mammalian predators; the bird was probably evolving towards flightlessness.

Blackbird

The sound of Europe

Māori name	Latin name	New Zealand status	Conservation status
Manu pango	***Turdus merula***	**Introduced**	**Naturalised**

The Eurasian or common blackbird holds the title of New Zealand's most widespread bird: whether you're in a suburban garden, on a farm or an island, or deep in native forest in isolated reaches of the country, you're still likely to see a blackbird. The male has the trademark glossy black coat with orange bill and eye ring, while the female is brown with a mottled belly (as are young birds).

Blackbirds were introduced to New Zealand from the 1860s, first with a shipment of just 26 birds by the Nelson Acclimatisation Society, and then by other regional societies. They were brought over to control insects like snails and slugs, and for a piece of nostalgic English birdsong. The song in demand was the male's: he utters his loud, fluting and warbling territorial song from a high perch from July to January, following up with a softer, more low-key song in late summer, then falling silent in autumn while he moults.

But then — *boom* — populations exploded. Within 15 years these adaptable birds were a pest, damaging fruit in orchards and spreading the seeds of blackberry, and within 60 years they could also be found on subantarctic islands, and the Kermadec and Chatham islands

Sometimes, if a blackbird spots a smaller bird with something tasty, it will rush it to drive it away.

as well. They turn up in about 90 per cent of gardens, both urban and rural, from sea level to about 1500 m altitude, but are slightly scarcer in South Island gardens, especially in Tasman and Otago.

Blackbirds eat mainly insects, spiders, snails, slugs and worms. They hop and run over the ground, stopping and starting, and cocking their head to listen for invertebrates or spy the tips of earthworms. They will also scratch about in leaf litter with their beaks and feet — mulch thrown off garden borders is usually the work of the blackbird. Sometimes, if a blackbird spots a smaller bird with something tasty, it will rush it to drive it away. Its fruit-eating habit makes the blackbird a major spreader of invasive weeds. Once eaten, a seed takes about 30 minutes to pass through the digestive

system, and the average blackbird will disperse seeds some 50–100 m away.

Blackbirds can be helpful, however, spreading native seeds such as coprosma, and they pollinate feijoa (originally from South America and one of New Zealand's best-loved tree crops). Along with mynas and thrushes, they eat the juicy feijoa petals (which can have sugar levels as high as nectar), collecting pollen on their heads and chests and spreading it to the next tree. Even on a self-fertile tree, the crop is always bigger when cross-pollinated with the help of the birds.

While the blackbird is a huge part of European culture — it is Sweden's national bird, and of course features in 'Sing a song of sixpence' — New Zealand's blackbirds have by now forgotten their European ways. Back in the UK, blackbirds are

Blackbird
in Ōtari–
Wilton's Bush,
Wellington.

LEFT
By chance, a mirrored window in a doctor's waiting room made the perfect hide from which to photograph this blackbird nest. Blackbirds are very sensitive to disturbance at the nest and will abandon it if they feel unsafe.

RIGHT
The distinctive orange beak and eye ring of a male blackbird.

parasitised by the European cuckoo, which lays its eggs in their nests, and so they will attack the adult cuckoo on sight. While New Zealand blackbirds will still eject eggs that don't look like their own, they will no longer attack a European cuckoo (or at least a taxidermied one in a study, presented at their nest).

Blackbirds make their cup-shaped nests in forks of hedges or shrubs. They return every year to the same territory, which the male defends — especially against other blackbirds that have a brighter orange or yellow beak. If it escalates to a fight, the birds sit facing each other with drooping wings, arched backs and lowered tails, then scuffle with beak and claw.

Blackbirds use all sorts of nesting materials — twigs, grass, moss, mud or rubbish, such as pieces of rope or wire mesh. A suspected blackbird nest found in Porirua in 2010 was thickly lined with the cotton-like cellulose acetate filters from cigarette butts (the toxins from the tobacco may have had the unexpected advantage of killing off nest parasites, but the effects of these virtual 'nicotine patches' on adult birds and blind, naked chicks is unknown). Females can lay three or more clutches of freckled blue-green eggs per season; the young are fledged and out of the nest within a couple of weeks, although they still take food from their parents for a few weeks more.

Brown quail

Australian replacement for an extinct local quail?

Māori name	Latin name	New Zealand status	Conservation status
Kuera	*Coturnix ypsilophora*	Introduced	Naturalised

At first glance, you could mistake the brown quail for an extremely plump and glossy brown rat as it scurries busily across the ground searching for food. But look closer and you'll see the plumage is exquisitely patterned with white streaks and black and brown chevrons. It's the perfect invisibility cloak for the scrub and rough grassland where these birds are most often found.

Hailing from Australia, the brown quail was introduced as a game bird to Auckland, Wellington, Otago, and Southland during the 1860s-70s. The only places it thrived, however, were the upper North Island (Coromandel, East Cape, North Auckland and Northland) and some offshore islands, such as Whale Island/ Moutohorā and some in the Hauraki Gulf. It is also found throughout Papua New Guinea and on the Lesser Sunda Islands of Indonesia, and was introduced to Fiji. Brown quail are often seen on the damp edges of wetlands and grassy roadsides.

They're voracious eaters of fallen seeds from grasses and shrubs, but also eat leaves and flowers, and pick invertebrates from foliage and leaf litter.

When frightened, a brown quail's first instinct is to squat in undergrowth or run for cover, but a covey (flock) of quail can also explode into the air, the birds splitting up in all directions, rapidly whirring their wings in low flight to land in a better hiding place. They fly well enough for such a dumpy-looking bird. For example, they self-colonised Tiritiri Matangi Island (now 200-300 birds), which is 3.5 km east of the

Whangaparaoa Peninsula — although this isn't a patch on their European cousins, which migrate between the UK and Africa every year.

Australian brown quail pair up in spring, usually nesting in long grass: they find a small hollow in the ground and line it with dead grasses. Then it's time to produce an enormous clutch of speckled white eggs: the hen lays one every day until she has as many as 12, and she does the lion's share of incubation. When they hatch, the tiny, downy chicks are the size of ping-pong balls and weigh only 8 g, and are very vulnerable to predators and bad weather. Nonetheless, they're ready to go exploring straight away and are taken on feeding missions to get bugs. To round up the troops, the parents give a penetrating shrill call. After a couple of weeks, the chicks can already fly, and in just a month they are as big as their parents.

After the summer breeding season, quail start forming coveys of up to 30 birds. When no longer nesting, they adopt their bizarre roosting habit of forming a perfect circle on the ground, heads to the outside, while they sleep — this possibly keeps them warm and lets them look out for predators at the same time.

New Zealand once had its own native quail, called koreke by Māori. It was widespread throughout both main islands, and numbers soared in the grasslands that sprang up after forests were burnt by Māori, especially in the South Island. To catch quail, Māori would find the little tracks the birds made through the bush, push a peg into the earth either side, and connect them with a low-hung string with slip-nooses attached. When Europeans/Pākehā arrived, they blasted the quail with lead shot, as they did with so many other New Zealand birds. Naturalist Walter Buller wrote that the first settlers 'enjoyed some excellent Quail-shooting for several years', citing the example of two men who in 1848 shot 43 pairs in a single day near Nelson. Other new threats to its survival included introduced predators like the Norway rat, and the conversion of grassland to agriculture by Pākehā. The koreke died out within about two decades, and by the mid-1870s had become one of the first in the wave of bird extinctions that followed Pākehā settlement.

Other quail were also introduced. California quail, found all over the country with its jaunty head plume, was introduced from 1865, and by 1890 became so numerous that in Nelson birds were canned or frozen and sent to London. (This was before the release of mustelids, which have decimated so many birds.) Also introduced was the bobwhite quail, which is probably now extinct in New Zealand.

Tiritiri Matangi Island in the Hauraki Gulf is an excellent place to spot these shy, wary birds as they venture across tracks.

Brown skua

Misunderstood assassin of the south seas

Māori name	Latin name	New Zealand status	Conservation status
Hākoakoa	*Catharacta antarctica*	Native	Naturally uncommon

The brown or subantarctic skua is a notorious ocean bully. This stocky, dark brown bird with a 1.5 m wingspan employs its stabbing beak to harass penguins away from their chicks before dragging a fluffy chick away to rip it apart; other times it'll mercilessly steal precious eggs for a protein-rich snack.

Fierce and fast, skuas can reach 80 km/h, readily dive-bombing anyone approaching their nest — and if you're an explorer or a researcher, 2 kg of feathers and webbed feet hitting at full speed is not something to be ignored. The heaviest skuas in the world, brown skuas are one of their region's top avian predators and may seem ruthless and cruel, especially to farmers on the Chatham Islands who see them feeding on dead or weak lambs, or to tourists in the subantarctic who watch them attacking newly hatched penguin chicks. They've been called the 'berserker among birds'.

Skuas have their warm fuzzy side, of course; they're extremely caring parents of adorable fluffy grey chicks that also must eat. They have always been a vital part of the ecosystem, cleaning up carrion such as placenta and dead pups of sea lions. They'll also pick off unfortunate seabirds on the subantarctic and Chatham islands (there are millions of seabirds in some skua-inhabited islands such as Rangatira in the Chathams, including 840,000 white-faced storm petrels and 330,000 broad-billed prions: food galore for the skua).

There are five species of skua seen

Their display pose, with wings held above, can seem almost demonic.

around New Zealand waters, but the brown skua alone breeds in this country. The largest local breeding populations are in the southern Chatham Islands — where the birds are called 'sea hens' by fishermen — and on the Snares Islands, but they also breed in southern Fiordland and on Stewart Island/Rakiura, as well as the Antipodes, Auckland and Campbell islands. Their nest is a scrape in the ground, sometimes lined with moss or grass, and usually in exposed areas near the coast. Pairs defend their territory fiercely from other skuas — their display pose, with wings held above, can seem almost demonic — and they hold the same territory for many years until they die or are ousted by a stronger bird. Unpaired skuas must hang out on 'club sites': allocated areas where no breeding territories have been set up. On the Chathams, skuas have benefited from the clearing of land for farming, which has provided more open area, but as the island reserves regenerate there may be less space in future for nesting territories. In the Chathams' relatively temperate climate, some birds will stay around for most of the year — but those that breed further south escape out to sea for the winter.

Some breeding skuas are comfortable with open relationships. Cooperative breeding, when birds get together in groups of three or more to raise chicks, is especially rare in seabirds, but brown skuas, especially those on the Chatham and Snares islands, seem particularly partial to it, with around 10 per cent of the population staying together in threes or larger groups for years (in the past, up to a third of nests operated this way). Usually, groups are one lady and two

Skuas keep a watchful eye over penguin and sea lion colonies, on the lookout for an easy meal.

unrelated gentlemen, but on one observed occasion two females and a male created a 'superclutch' of eggs.

A skua doen't turn its formidable beak up at much. As well as penguin chicks and eggs (which, to be fair, are actually quite hard to get — a mob of penguin adults can do some serious damage to a skua), it'll grab small adult petrels and prions on the ground at night time or pre-dawn, snatching them as they scramble across the ground to and from their burrows. It will even go for a bird as large as the black-backed gull. Skuas also hang out round sea lion and fur seal colonies, where there's nothing better than a stinky adult carcass, a dead pup or a piece of nutritious placenta, fresh umbilical cord or even dung. They'll also eat goose barnacles and small marine invertebrates. Out at sea, brown skuas take prey at or just below the sea surface, or even kill other seabirds by forcing them underwater again after they've plunge-dived.

Smaller skua species carry out kleptoparasitism, looting what someone else has caught. They get other birds to disgorge or drop their food mid-flight by chasing them, giving ear-shredding screams and dropping down on them in the air, after which they scoop up the forfeited food as it falls. In the past, it was thought the skuas were eating falling bird poo — which in some circles earned them the name *Stercorarius*, Latin for 'of dung'.

Numbers are not clear, but the brown skua is thought to be in decline, with about 26,000 left. Skuas are occasionally shot: they are only partially protected under the Wildlife Act, and if they are causing injury to property they may be killed by the landowner or occupier.

Fluffy and grey with big paddle-feet, skua chicks are well guarded by their parents when small.

Canada goose

A force to be reckoned with

Māori name	Latin name	New Zealand status	Conservation status
Kuihi	*Branta canadensis*	Introduced	Naturalised

You've got to hand it to the Canada goose. This big introduced bird with a chinstrap could give all other birds a lesson in adaptability, population growth and robustness, having successfully taken over paddocks and waterways around the country — and, in fact, the world.

In the high latitudes of its native North America this goose is virtually the ironman of the bird world. It can migrate 1600 km or more, clocking 850 km in less than eight hours nonstop. It does this by flying in a classic 'V' formation: research has shown 25 birds flying in a V have 71 per cent more range than if they were to fly solo (it's the upwash from each bird's wings that benefits the birds beside it). In New Zealand's mild climate, however, there's no need for the goose to migrate; instead, it's more of a fun-run athlete, flying only short distances between lakes and grasslands to breed, honking exuberantly all the way.

There are two species of Canada goose, the larger of which — *Branta canadensis* — is found in New Zealand. It was introduced as a game bird, in 1905 and again in 1920, but it is an expert evader of hunters: it feeds in open areas, remaining alert and warning the rest of the gaggle of danger, outwitting most hunters unless they use camouflage or decoys. By 2010, numbers had exploded: there were 60,000 Canada geese in New Zealand — two-thirds in the South Island. These successful newcomers were wreaking

LEFT AND RIGHT Sweet, tiny goslings take 8–9 weeks to transform into smartly-patterned adults.

havoc: four or five birds will eat as much grass as one sheep. They were gobbling up the aquatic plants that the native waterfowl love, and fouling paddocks and waterways with their copious amounts of large green-and-white droppings. Crowds of Canada geese can plague airfields and in some countries have caused major incidents when they've collided with planes. The 2009 crash-landing of US Airways Flight 1549 into the Hudson River was attributed to a flock of migratory Canada geese sucked into the plane's engines.

Accordingly, New Zealand's Canada geese lost their protection in 2010 and can be hunted as pests at any time of the year, without a licence from Fish & Game.

Indeed, this is the only game bird managed in order to keep numbers *down*, not up. Culls are organised by local communities when necessary. In areas where the geese are a constant problem, such as Molesworth Station and Christchurch International Airport, hunting groups run annual culls as part of a bird management programme. The process is sometimes more efficient when the birds gather to moult in December–January: at this time, having lost both their primary and secondary wing feathers, they become flightless for about a month.

Canada geese live for about 25 years and bond for life; they can breed from when they're two years old, right through to 18. Males in particular are extremely

aggressive around nesting time, protecting their brooding mate and eggs or chicks — they'll launch an attack at any threat, charging with flapping wings and outstretched neck, giving a nasty peck. The yellow-grey goslings imprint on their mother or first carer. The 1996 movie *Fly Away Home*, starring Anna Paquin — in which Canada geese imprinted on her and she taught them a migratory route by having them follow her biplane — was based on the true story of Ontario man Bill Lishman, who began training geese to follow his ultralight aircraft in 1986, leading them on a migration in 1993.

Within 48 hours of hatching, the pint-sized goslings are able to travel 1.5 km on foot or 14 km on water. If there are many nests close together, goslings can wander off and join other broods, forming 'gang broods', where between eight and 60 goslings wander about together with one or two adults, who have been landed the mammoth task of watching over them. The young stay with their family group for several months.

Before the arrival of humans, New Zealand had its own native goose, a flightless bird that inhabited open grasslands on both main islands. The South Island goose was the largest: weighing 18 kg and standing a metre tall, it was about three times the size of a Canada goose.

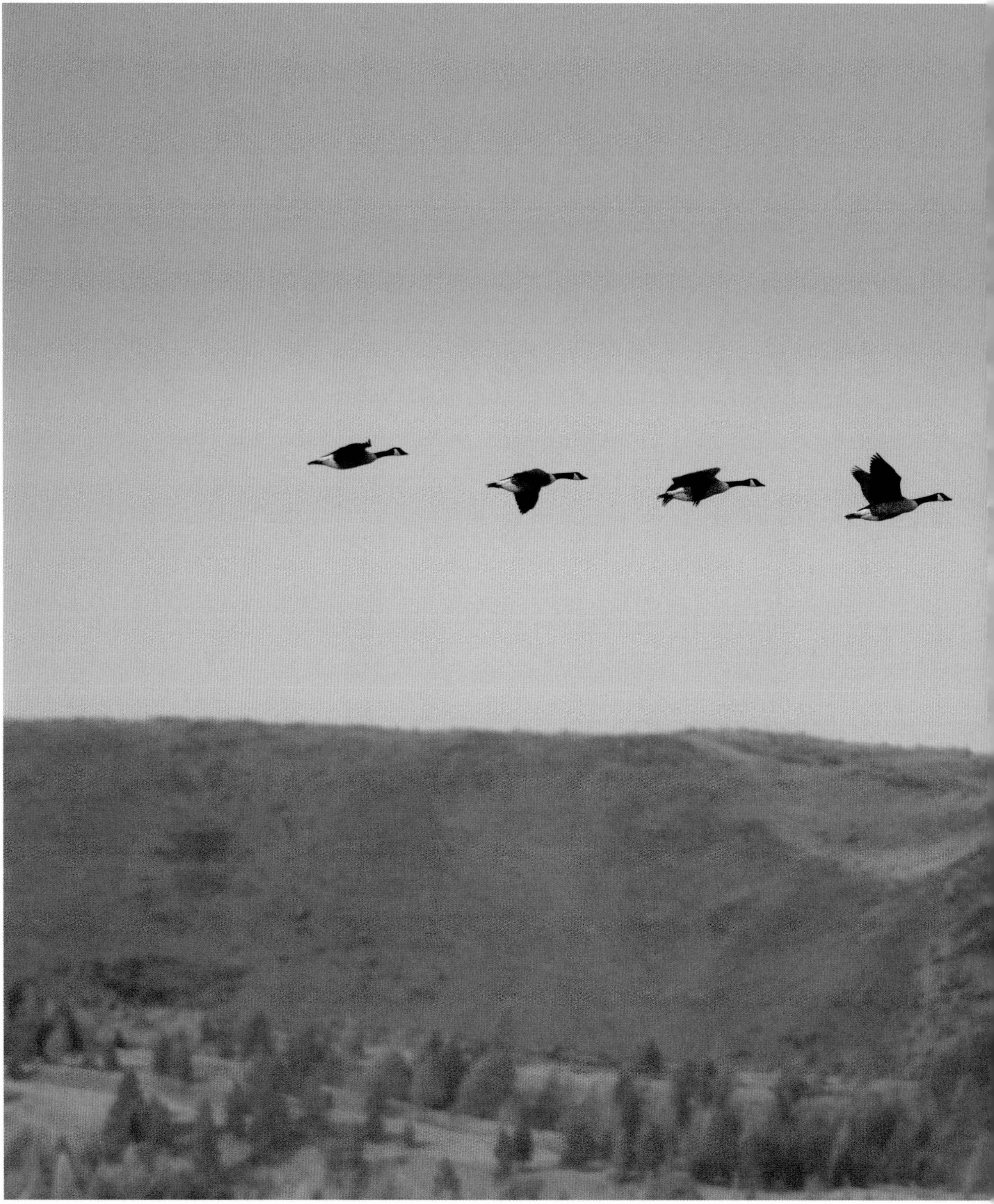

Formation-flying over the Tasman River delta.

Common myna

The skills to thrive internationally

Māori name	Latin name	New Zealand status	Conservation status
Maina	*Acridotheres tristis*	Introduced	Naturalised

This bird with the yellow racer sunglasses, beak and stockings — paired with a brown coat with white flashes — has excelled on a global scale. The common or Indian myna is so good at surviving that it has become notorious, outcompeting and displacing native species on all continents save South America and Antarctica.

The myna is rated among the world's top 100 worst invasive species — one of only three birds on the International Union for Conversation of Nature list along with the common starling and the red-vented bulbul. It is a pest in New Zealand, South Africa, North America, the Middle East, many Pacific islands and especially Australia, where it is sometimes called the flying rat or flying cane toad (the neotropical cane toad is another deeply unwelcome import there). In Fiji, where the myna was introduced in 1890 to control sugarcane pests, it damages all manner of crops.

How did the myna acquire such a villainous status? For starters, it is aggressive. Just slightly bigger than a blackbird, it has refined the art of intimidation by mobbing and swooping and threatening other birds in their territories. Studies have shown that when mynas are removed, native bird presence increases greatly. If the myna needs a

If they see another myna captured by a human (and presumably by other predators) they'll remember it and be wary from then on.

nest, it can just steal one — aggressively evicting the adults, even grabbing the chicks and dropping them to the ground from high up, and destroying eggs. It does this to saddlebacks (it steals their specially constructed nest boxes on Tiritiri Matangi, for example), starlings, tūī, kingfishers, New Zealand robins, kōkako and many others — sometimes, it doesn't even use the nest, acting simply to intimidate.

Another way to survive out there in the world is to be resourceful: eat anything. Mynas are omnivorous. They can eat snails, beetles, worms, flies and all manner of other bugs, plucking them from fields or grain stubble. They take the tiny carcasses of insects hit by cars on roadsides — and the larger, meatier carrion too, for that matter. They'll devour birds' eggs, chicks

and lizards. They'll raid orchards for fruit, and damage grain crops like maize, wheat and rice. Scavenging at local dumps is also a favourite tactic, as is stealing food morsels outside cafes. In coastal areas, mynas turn their attention to crustaceans, marine worms and molluscs at beaches, mudflats and rockpools.

These birds are also extremely smart. If they see another myna captured by a human (and presumably by other predators) they'll remember it and be wary from then on. All of this makes the myna a very successful invasive pest that is hard to control by shooting or trapping.

A member of the starling family, the common myna is native across Asia, from Afghanistan through India to Bangladesh. Several hundred were introduced to parts

An over-abundance of mynas can lead to fruit crops being decimated.

of New Zealand during the 1870s by acclimatisation societies and individuals to help control insect pests (in India the myna is known as 'the farmer's friend' for its appetite for insects). They have thrived in New Zealand north of the Manawatū — except for areas of the volcanic plateau and the dense forest of Te Urewera — and are common in farmland, parks, gardens, orchards and forest edges. Mynas didn't push north of the Waikato until the 1950s, but when they did, their numbers exploded in the warmer climate and they've made themselves completely at home ever since.

A related species, the hill myna, is a popular pet in many countries for its incredible skill in mimicking human speech, and its capture for this purpose has led to local declines in some wild populations. The common myna didn't inherit this skill; its cries are raucous whistles, gurgles, growls, chatters and bell-like notes. And the common mynas introduced around the world are even more basic in song than the common mynas in Asia: a recent study showed all introduced populations have lost complexity in their song. This is possibly

Mynas thrive in modified
suburban and urban
habitats, such as central
Auckland.

a consequence of the 'founder effect' — the loss of genetic diversity that results when a large introduced population grows from a very small number of individuals.

Regardless, you'll hear this cacophony often: mynas are extremely social and roost together, from just a few birds to thousands, often in isolated stands of trees, such as pōhutukawa or macrocarpa, or under the eaves of buildings. They chatter raucously before dawn, and in the breeding season the first thing the male does is head for his territory (the female will have been on the nest all night incubating eggs or nestlings) and start proclaiming it's his to the world, singing loudly for up to 15 minutes and bobbing his head. Aggressive fights break out, especially at territorial borders, with birds pecking at each other, jabbering loudly and wrestling on the ground — it's always males against males and females against females, and usually fights are over territories, food or roost sites. Mynas will also swoop at any cat, dog or human they perceive as a threat to their nest. Outside the breeding season, they are often seen in flocks during the day.

Mynas pair for life, and in India this has made them a symbol of everlasting love.

Every year they hold the same territory (0.83 hectares on average, which is more than 10 regular-sized suburban gardens), and raise two broods of up to six chicks each. If they haven't stolen some other bird's nest, they'll make their own — in tree cavities or dense vegetation, but also often in drainpipes and spouting, much to the annoyance of homeowners. They use whatever they can find: grass, sticks, paper (and even snake skin in other countries), lining the nest with soft green leaves, fur, feathers or soft plastic.

Mynas will sometimes form a pair when they're less than a year old, but usually only breed when a little older. They live for about four years in the wild, with some in New Zealand reaching the grand age of 12.

Mynas have a terrible reputation, but they have the sass and gumption that it takes to survive in a harsh world. Their scientific name *tristis* means 'sad' or 'gloomy', which refers to their dull chocolate-brown coat, but this bird arguably doesn't have too much to be gloomy about at all.

Common starling

Murmurating mimic

Māori names	Latin name	New Zealand status	Conservation status
Tāringi, whātete	*Sturnus vulgaris*	Introduced	Naturalised

'A drunken fingerprint across the sky'
— that was how American poet Richard
Wilbur described a murmuration of
common starlings: the swarm of hundreds
or thousands of birds, swirling through the
skies before they settle on their roosts at
night during winter. While they're certainly
not drunk, how they do this has sparked
myriad studies from which there are many
theories, ranging from keeping an eye on the
flight angle of their neighbours, to watching
the patterns of light and dark around them,
to a 1930s theory of telepathy (less likely).
The Romans considered the murmuration
patterns to be messages from the gods.

The young are brown, but the adults develop a sumptuous winter coat.

However they do it, starlings are famous around the world for this display — which is thought to be a 'safety in numbers' ploy to avoid birds of prey, a bit like shoaling in fish. They have a few other tricks up their sleeves, too. They are incredible mimics, and are one of the most successful introduced species in the world: they are fully at home now (to various degrees of invasiveness) in the USA, Australia, South Africa, Argentina and many other countries, including New Zealand.

On the ground, they walk rather than hop; it's almost like a swagger, and they're very good looking. The young are brown, but the adults develop a sumptuous winter coat. In the sun they shine with a gorgeous iridescent purple and green with the cream tips of the feathers creating a flecked effect — and these 'stars' disappear as the feathers wear.

This bird has infiltrated every part of New Zealand and its islands apart from alpine areas and the deepest native bush. It is the pasture critter eater extraordinaire: earthworms, caterpillars, beetles and spiders. Starlings can feed in large flocks sometimes of several hundred birds, each jabbing the ground and widening little holes with its beak — the flock can look like it's rolling as the birds leapfrog each other. They're not fussy, though, also scavenging food scraps from cafes and gardens, pecking out holes in soft fruits in orchards, and sampling nectar from flowers.

About 1000 of these murmurating characters were introduced to New Zealand from Europe in the 1860s, by acclimatisation societies and private individuals, in a bid to control the insect plagues that devastated crops and pastures

In their annual moult, starlings trade oily black plumage for gold-tipped iridescence.

on deforested land. But once introduced, their numbers exploded. Soon, they were damaging cereal crops and orchard fruit — vineyards too — and in the autumn their feeding flocks can strip the small fruits from native trees, such as kahikatea, removing valuable food for tūī, kererū and bellbirds. Starlings are also involved in a large percentage of the aircraft strikes that happen globally every year. In 1960, a flock of starlings was responsible for a plane crash in Boston that killed 62 people — the birds were sucked into multiple engines not long after take-off, and had also obscured the pilot's vision.

Starling males begin nesting at two to three years; females at just one to two. They nest in cavities and their sites are diverse: tall stands of pōhutukawa or macrocarpa, farm machinery and chimneys, under eaves. They also take over nest boxes meant for endangered birds. They fill their chosen spot with dried grass and whatever else is available. Uniquely in the New Zealand population, they can also excavate clay banks to nest.

The female incubates the pale blue eggs through the night, and the male relieves

Starling calls vary from raucous song to pops and crackles and electronic-sounding buzzes.

her for the morning. Instead of defending a whole breeding and feeding territory like many other birds, the male defends the nest site only, staying nearby all day. He sings a territorial song from a perch up high, with lots of whistles, clicks, warbles and gurgles. Starlings are noisy birds in general, and they're a very good mimic of other bird calls and neighbourhood sounds, like police sirens. They can be taught to mimic human speech, too, speaking with an eerie, computer-like tone. Historically, they were a widely kept pet in Europe, and Mozart famously kept one that he had overheard singing his music in a neighbouring pet shop.

Males and non-incubating females roost together in their noisy thousands, flying up to 30 km every day to join their cronies in a murmuration and a cosy communal night's sleep. Often they roost in favoured trees on islands. New Zealand, with all its predator-free islands, is great for starlings, but starlings can spread avian diseases to vulnerable native bird populations, and also the seeds of weeds, such as boxthorn. In cities, they roost in trees but also in sheltered areas such as the eaves of railway or bus station buildings, much to the dismay of anyone who parks underneath:

the corrosive power of the acidic droppings of thousands of roosting birds is not to be underestimated. From all around the globe come frequent reports of the astronomical cost of cleaning up starling poo or installing scarers; Palmerston North City Council alone spent $50,000 on bird scarers for its central square in 2017.

Starling numbers plunged in New Zealand in the 1950s when DDT was heavily used as a pesticide on farms and orchards. Use of the chemical was banned in 1989, but it still circulates in soils and, probably because they eat so many earthworms, New Zealand's starlings can still have more DDT in their eggs than starlings in any other country in the world. However, the levels no longer seem to be toxic as they are still one of our most common birds.

Dotterels

Fat little scurriers with come-hither bibs

Northern New Zealand dotterel

Māori name	Latin name	New Zealand status	Conservation status
Tūturiwhatu	*Charadrius obscurus aquilonius*	Endemic	Nationally vulnerable

Southern New Zealand dotterel

Māori name	Latin name	New Zealand status	Conservation status
Tūturiwhatu	*Charadrius obscurus obscurus*	Endemic	Nationally critical

Banded dotterel

Māori names	Latin name	New Zealand status	Conservation status
Tūturiwhatu, pohowera	*Charadrius bicinctus*	Endemic	Nationally vulnerable

Black-fronted dotterel

Māori name	Latin name	New Zealand status	Conservation status
n/a	*Elseyornis melanops*	Native	Naturally uncommon

Northern New Zealand dotterels are a common sight on beaches in the Bay of Islands.

These pudgy little birds dash to and fro across wetland margins and beaches — run, stop, peck is the name of their game — and some can also be heard in pebbly rivers or high country, with their contact call of *chip chip*, accelerating when in danger. New Zealand's dotterels come in four flavours: northern New Zealand dotterel, southern New Zealand dotterel, banded dotterel and, since the 1950s, black-fronted dotterel. The first three have a white chest that turns a romantic rusty red in the breeding season, and they are also in varying degrees of trouble.

Northern New Zealand dotterel

The northern New Zealand dotterel is our beach babe, laying camouflaged eggs in sandy or pebbly scrapes just above the high tide mark on beaches and sandspits mainly from North Cape to East Cape. It patrols the shore to snap up bivalves, small fish and a wide range of terrestrial and marine invertebrates; grinding up shell is no problem for the hardy digestive system of this bird. Its favourite food in tidal estuaries is crab, which it gruesomely dismembers. Dotterel expert John Dowding describes the bird's habit of grabbing the crab by a leg and shaking it violently until the leg comes off. As each leg is eaten, the poor crab loses its mobility, and eventually all that remains is the torso, which the dotterel then gulps down whole — the bulge can be seen sliding down its throat. Foot-trembling is another ingenious way to get food: the bird trembles one foot on the surface, cocks its head and stabs at anything that moves. Both northern and southern New Zealand dotterels use this trick to get sandhoppers

Mudflats are a buffet of marine food for New Zealand dotterels.

out of mounds of seaweed or sand, and invertebrates from shallow sandy pools.

When they nest, beach- and riverbed-nesting dotterels blend perfectly with sand and pebbles: their dark back feathers have pale fringes, giving a fish-scale appearance, and the eggs are mottled. But the poor old northern New Zealand dotterel and banded dotterel struggle to find a safe spot on narrow beaches — their nests often have to be relocated by human volunteers before the tide washes them away. There are perhaps 2500 northern New Zealand dotterels, and while their number has been gradually increasing, these birds depend wholly on humans for their survival. Without the efforts of conservation groups (predator control in breeding season, shifting their nests away from high tides and roping off their territories from disturbance) their numbers drop. Curiously, about 10 per cent of northern New Zealand dotterels in Auckland have started breeding further inland — something this subspecies didn't do previously — on grassy paddocks, building sites, motorway verges and even undeveloped land in the Albany Mega Centre. An explanation is the enormous pressure an expanding city puts on its

coasts; gone are many of the dune areas in urban Auckland, replaced with the retaining walls of waterfront sections and roads. One exception is Shoal Bay on the North Shore — a group of birds there is thriving on grassed areas and artificially raised shell pads next to a humming motorway, which acts as a barrier to predators. When predators — which include cats, rats, stoats and hedgehogs, as well as larger birds — do get to a dotterel nest, adult birds will feign injury to lead them away, and the braver birds will even do their best to fly at them, squealing, but unfortunately the predators are little deterred.

Southern New Zealand dotterel

The southern New Zealand dotterel, a heavier and darker bird, is a rugged high-country ranger that is now extremely endangered. It disappeared from the mainland by 1900, and there are now only 140 birds left, breeding only on the subalpine herbfields of Stewart Island/ Rakiura. Conservation efforts increased the population to 300 in the 1990s, and it's not clear why they've recently collapsed,

although cat and rat predation could be a factor. The chicks hatching on these hilltops have a completely different diet to their North Island counterparts, eating worms, spiders and insects. When free of incubating duties, adults slip away for better food: they fly down to intertidal flats — where they are found during the rest of the year — and get their fix of fat bivalves, marine worms and fish.

Female dotterels occasionally couple up, more so when males are scarce. Generally, males are more likely than females to be killed by nocturnal predators, because the male usually incubates at night while the female is out feeding. This can lead to a sex imbalance, particularly in the small southern New Zealand dotterel population. When females can't find a male, they can set up a territory and nest with another female, together laying double the number of eggs. Sometimes, one of the eggs will even miraculously hatch — researchers aren't sure if this chick is a lovechild of a cheating paired male, or the product of a wandering male not ready to settle down. However, while male-female pairs rarely divorce,

Banded dotterels also breed in high-country braided river habitats, where they are well camouflaged amongst the river stones.

female-female bonds in dotterels last only a season or two — generally, as soon as an eligible male turns up, he'll snaffle one of them.

Banded dotterel

Banded dotterels — named for their double breast bands — are New Zealand's most common dotterel at about 20,000 birds, but they are declining. They are found the length of the country and on offshore islands, on shores, estuaries, riverbeds and in the high coutry. They eat anything from crustaceans and worms to spiders and insects, but there's no dinner conversation — the banded dotterels always forage alone, even in the breeding season, often eating at night on estuaries. They will, however, flock when roosting, settling on islets or bars formed by the high tide.

Black-fronted dotterel

The black-fronted dotterel came over from Australia and started breeding in New Zealand in the 1950s. It's smaller and slimmer than the other dotterels, and is still relatively uncommon. It tends to avoid the coast, sticking to lakes, ponds, rivers and brackish estuaries.

While they happily feed close together in the intertidal sands and mudflats, dotterels squabble over territory boundaries that are close to the nest, fighting with their beaks. Pairs keep their territories for years, the northern and southern New Zealand dotterels leaving them in January for only a few months to join a post-breeding flock, which may be up to 30 km away. It's here that birds meet their partners — whether they're finding love for the first time or new love after loss. Banded dotterels are more likely to migrate; some, such as those that breed in the Mackenzie Country, escape the inhospitable winter by migrating all the way to south-eastern Australia — a journey of at least 1600 km! Coast-breeding banded dotterels don't go to such extreme lengths, merely heading to northern New Zealand harbours.

Ducks

More intrepid than your average bread-eating park-dweller

Blue duck

Māori names	Latin name	New Zealand status	Conservation status
Whio, korowhio, kōwhiowhio	*Hymenolaimus malacorhynchos*	Endemic	Nationally vulnerable

Paradise shelduck

Māori name	Latin name	New Zealand status	Conservation status
Pūtangitangi	*Tadorna variegata*	Endemic	Not threatened

New Zealand's ducks have blazed their own trails, each one perfectly adapted to a special slice of environment — from the blue duck or whio, which has perfected the art of scraping caddisfly larvae from rocks in foaming torrents of rocky rivers, to the scaup that dives deep in lakes for most of its food, to the swamp-loving pāteke or brown teal. Some, like the whio, are barely clinging to survival while others, such as the paradise shelduck, are thriving.

Blue duck

The endemic blue duck or whio is at home in the white water of remote forested headwaters. The male's piercing *whio* whistle accounts for the Māori name; the female, in contrast, utters a low-pitched rasp, a bit like a creaking door. This duck is slate blue with a bright yellow eye: its grey webbed feet are whoppers, providing the paddle power to scoot through powerful rapids; the webbing closes on the forward

Made for rapids and white water, these birds have no difficulty navigating swift alpine rivers.

stroke to create less drag. Large, strong claws help the duck scrabble onto rocks. It's born for this: even the fluffy, pint-sized chicks will bravely get into the white water.

One of the strangest things about the whio is its bill. This is pink, with dark, fleshy and rubbery flaps attached to the tip which it uses to feel for invertebrates on rocks and scrape them off. So the attachments are a bit like a lip, but they also protect the bill from being worn away. Inside the bill are sieve-like structures to filter out prey. It feeds this way at dawn and dusk, roosting out of sight for most of the day.

Pairs usually set up a territory not far from where they hatched; they will hold it year after year, and defend it fiercely. They nest and roost in riverbank vegetation, their chestnut-coloured chests providing camouflage. Some birds pair for life, and while others switch partners after a few years, the males are nowhere near as sexually promiscuous as, say, mallards and muscovies.

Whio thrive only in the cleanest of rivers, making their appearance on your local river a bit of a gold star in terms of water quality. They also need the river channel to be stable, with a deep, narrow centre and shallow edges, with forested banks. Although they can fly long distances overland when they need to, generally whio prefer to swim, or just fly very low, following the river's course.

Before the arrival of humans, whio were everywhere from alpine tarns to lowlands. Today they are in trouble, mainly because of introduced predators, especially stoats. Island sanctuaries offer no refuge because these ducks need fast-flowing rivers. Riverside predator control is practised on some rivers, such as the Wangapeka and Fyfe rivers in Kahurangi National Park, the Tongariro River in the Kaimanawa Ranges and on Mt Taranaki (where whio were once extinct and later reintroduced) — but in places where there is no such control the birds are in decline. Predators take around 50 per cent of ground-nesting females and 60 per cent of fledged young; up to 90 per cent of nests fail because of egg thieves, and the ducks rarely nest again in the same season.

The whistling of a male whio rings clear over the swift waters of the Tongariro River.

Paradise shelduck

In ducks, the male tends to be the more flamboyant, but paradise shelducks flout this rule — they could be called the cross-dressers of the pastures. The female has a gorgeous chestnut-brown body and a pure white head and neck, while the male is mostly a dark bird with only small patches of colour. They were called the 'painted duck' by James Cook, who saw them in Dusky Sound on his second voyage in 1773, and they are found throughout the North and South islands.

For eight or nine months of the year, pairs are practically inseparable. They usually bond for life (although if one dies, the other moves on to a new partner fairly quickly). The Māori name for the paradise shelduck is pūtangitangi — tangi means 'a lament' or 'weeping for the dead'. This is thought to be in reference to the sad, haunting wail of the female's high-pitched warning or flight call. By contrast, the male gives a goose-like honk (a shelduck is halfway between a duck and a goose).

Shelducks nest in the same territory every year. Usually they bed down in a small hollow on the ground lined with down and hidden under logs or tree roots, but sometimes they might nest 20 m high in trees. The female sits on the nest while the male defends the territory — and he'll accompany her as her bodyguard when she leaves the nest to feed. Each season, a single large clutch is laid, and incubated for a month. Just a day after hatching, while still unable to fly, the fluffy black-

Female paradise shelducks are more colourful than the males, and the sexes also have different calls.

and-white striped ducklings leave their nests — if in trees they plummet to the ground but miraculously are unharmed — and waddle along after their parents to water (sometimes more than a kilometre away). They eat aquatic insects to start with, moving on to plants after about two weeks.

Once the parents have finished raising their chicks, they'll join other adults in a moulting flock on a large body of water and lose all their flight feathers at once, rendering them flightless for three or four weeks. Forming large groups during the moult makes them less vulnerable to predators like harriers.

The pūtangitangi is one of the few native species that have benefited from the clearing of New Zealand's land for agriculture. They eat grasses, clover, grains, seeds and aquatic vegetation, and are found in pastures, cropfields, wetlands, rivers and lakes. They'll even become urban ducks, following waterways into towns and cities. Before forests were cleared, shelducks were not common: early Māori were careful to harvest them outside the breeding season only, while they were moulting, and even then only the fat ones were taken. Today, they are legal game in the hunting season, with about one-third of the population shot every year (200,000 of about 600–700,000). This would be a death sentence on many other native bird species, but with a large clutch of five to 15 eggs laid by the female each year, the paradise shelduck is thriving.

Eastern bar-tailed godwit

Obese, yet ironman-ready globetrotter

Māori names	Latin name	New Zealand status	Conservation status
Kuaka, kura	*Limosa lapponica baueri*	Native	Declining

The eastern bar-tailed godwit is a bird of legendary abilities. It pulls off the longest non-stop migratory flight of any bird so that it can have year-round access to the best food the world has to offer. Twice a year, it eats its way to obesity and then powers in constant flight at about 60 km/h from one hemisphere to the other — an incredible journey of more than 11,000 km between Alaska and New Zealand.

Out on New Zealand's mudflats it's a generic-looking bird, but for an unusually long beak, slightly curved upwards — the female's beak is even longer than the male's, over half her total body length. The godwit's usual outfit is of browns and greys, but before taking off for the breeding season in western Alaska the male really turns it on: his head, neck, breast and belly turn a rusty red and his patterned upper-part feathers turn up the contrast. Females gain a pale red flush and some strong streaks.

But inside this relatively plain bird an anatomical miracle takes place. Twice a year from the age of three or four — and once when it was about four months old — the godwit's body begins to almost remodel itself a few weeks before the journey between New Zealand and

Trading plain brown for ruddy breeding plumage, this male is preparing himself for a long flight north to Alaska.

Alaska. It needs to pack on enough weight to provide fuel to sustain it for more than a week. The bird eats gluttonously, sometimes even doubling its body weight, packing it all around the intestines, heart and gizzard, plus a thick layer under the skin. (The fattest godwit ever recorded was 55 per cent fat — that was in Alaska, before its migration back to New Zealand.) Then, since the godwit won't be able to eat during the long flight, it shrinks its digestive organs — which would otherwise burden it down with weight — absorbing the protein for fuel to burn en route. The bird maintains only the essentials for flight — wing muscles, heart, lungs — and legs to carry its blubbery body around before departure. In New Zealand, the godwit also undergoes its annual moult into fresh flight feathers that last the near-30,000 km round trip. It even changes components of its blood in order to burn fat more efficiently.

To pack on all that weight, a godwit constantly walks along the intertidal flats and probes at the sediment every few steps to find the single food that comprises up to 94 per cent of its diet: marine worms. Sensory organs at the tip of its pink-and-

black beak detect the vibrations of these worms — any hint of movement and it'll plunge its beak into the mud (sometimes up to its eyes); special muscles enable the tip of the bill to pincer prey even when buried deep in the mud. It'll also go for the odd mollusc and crustacean, and occasionally glean worms and other invertebrates from wet pasture.

In March, the godwits head to Alaska via a few weeks' stopover on the Yellow Sea coasts of China and Korea; 10,000 km the first leg, 6000 the second. In early September they get back to New Zealand — this time direct in eight or nine days; 11,000 km in one go. The March pre-flight bird has a plump chest and low-hanging, almost round stomach; the September post-flight godwit, which is scrawny with a sharp breastbone, is unrecognisable as the same bird. She has burned her fat and a lot of her muscle — the lighter she got in flight, the less muscle she needed. Her wings may droop with fatigue, and her flight feathers are tattered. And the first thing she does, apart from having a long drink, is sleep.

Godwits occur all around the world, and the main godwit in New Zealand is

the eastern bar-tailed godwit, subspecies *baueri*. While other subspecies might head to places like Cape Town and Thailand for the southern hemisphere summer, this one turns up in harbours and estuaries all around the New Zealand coast, with major congregations in Pārengarenga, the Firth of Thames, the Manukau and Kaipara harbours and Farewell Spit. About 40 per cent of the *baueri* subspecies will head to Australia instead.

Individual birds leave New Zealand for Alaska at about the same time every year. Early Māori didn't know where the godwits disappeared to, hence the proverb, Kua kite te kōhanga kuaka? (Who has seen the nest of the godwit?) In Northland, near Cape Rēinga, the departure of the godwits is linked with the departure of human souls from this world.

When the godwits get to western Alaska, they'll split away from the flock and head towards their pair territories, where the male starts his fantastic aerial courtship displays and nesting begins. They breed anywhere from coastal wetlands to the tundra 200 km inland.

Their world in the north of the globe is completely different to Aotearoa. They deal with Arctic foxes, ravens and long-

Flocks of northbound godwits are a common sight at the end of the austral summer.

tailed skuas. The Alaskan summer, with its long sunshine hours, buzzes into life after the thaw, with insects and vegetation springing up in huge abundance — just right for raising healthy chicks. The female lays three or four eggs in a shallow bowl lined with lichen; they are massive, each around 11 per cent of her body weight. This allows the chicks to hatch self-sufficient, ready to catch their own insect food, and fledging within 20 days. At just four months old, these young birds embark on their epic journey to New Zealand, sensing when conditions are right for a good tailwind.

The fact that godwits do a nonstop flight from Alaska to New Zealand was proven only in 2007 by American biologist Robert Gill, when he tracked a female known as E7 with a satellite tag. She left on 30 August and arrived 7 September, flying 11,680 km in just over eight days. Why do they fly nonstop south? Theories are that they avoid the predators and disease they could encounter by stopping over, and they also save time and energy. But by stopping via the Yellow Sea on the northbound journey, they can build up some more fat for three or four weeks in order to arrive in Alaska in good enough condition to court and breed, especially if food is low in the recently thawed land.

While these super-athletes would be terrible candidates for dinner after their harrowing journey, they were often eaten in the past when they were in fat condition

At just four months old, these young birds embark on their epic journey to New Zealand.

by both early Māori and early European settlers. One of the ways Māori would catch them was by laying out long flax nets hidden under a thin layer of sand on the sandbars at low tide. As the tide rose and the birds moved onto the sandbars, their feet would get tangled up in the net. Europeans would shoot the birds as they massed together to roost — and many would shoot them just for sport. Godwits were treated as a game bird on and off for many years and were protected finally in 1941, although illegal hunting continued for some time — and may still continue to this day.

Since 2016, the *baueri* subspecies has had a conservation status of At Risk, Declining, because its population of 150,000 is thought to be declining by around 2 per cent each year. One huge reason could be habitat loss at the birds' Yellow Sea stopover; in China and South Korea there is extensive reclamation of estuaries, and sea wall construction, for cities and fish farms. The birds lose their worm-rich intertidal flats and most roosting sites, and are exposed to a lot of pollution. Global warming could bring major changes for these birds — such as shrubs and trees colonising their tundra and encroaching on their breeding habitat, and disruption to wind patterns that would usually aid migration.

Their stopover on the coasts of China and South Korea might be shrinking, but most of the coast of North Korea is still

Godwits mass in their thousands at Miranda in the Firth of Thames.

undeveloped. Since 2009, a group from New Zealand, including godwit experts Adrian Riegen and Keith Woodley, has, incredibly, been allowed access to North Korea, to count and watch the birds stopping off on their northward migration. Since the 1990s, the group had been surveying the birds around South Korea and China, and noticed the birds flying into North Korea, but nothing was known about their numbers there. They wrote to politician Winston Peters, who was to undertake a diplomatic visit to the hermetic kingdom, and he managed to secure them rare permission to survey birds on highly sensitive parts of the coast — areas that still have the extensive reedbeds and mudflats lost elsewhere, and which are so essential for the godwits' refuelling. The awareness the group has brought to the North Korean government of the plight of the birds has led to the

North Koreans signing the Ramsar Convention on Wetlands of International Importance, and being accepted into the East Asian–Australian Flyway Partnership.

Other godwits that occasionally visit New Zealand are the black-tailed and Hudsonian godwits. The black-tailed godwit we see here breeds in northeast Asia and overwinters in tropical Asia and Australia. It has a straighter bill than the bar-tailed godwit, is slimmer with smoother grey plumage and has a black tail that is obvious in flight. It comes to New Zealand in tiny numbers — and it usually snubs the bar-tailed godwit: if it doesn't have any of its own kind, it prefers to hang out with pied stilts. The Hudsonian looks similar to a black-tailed godwit but has a blackish (not white) underwing. Most of its number head to South America, rather than New Zealand.

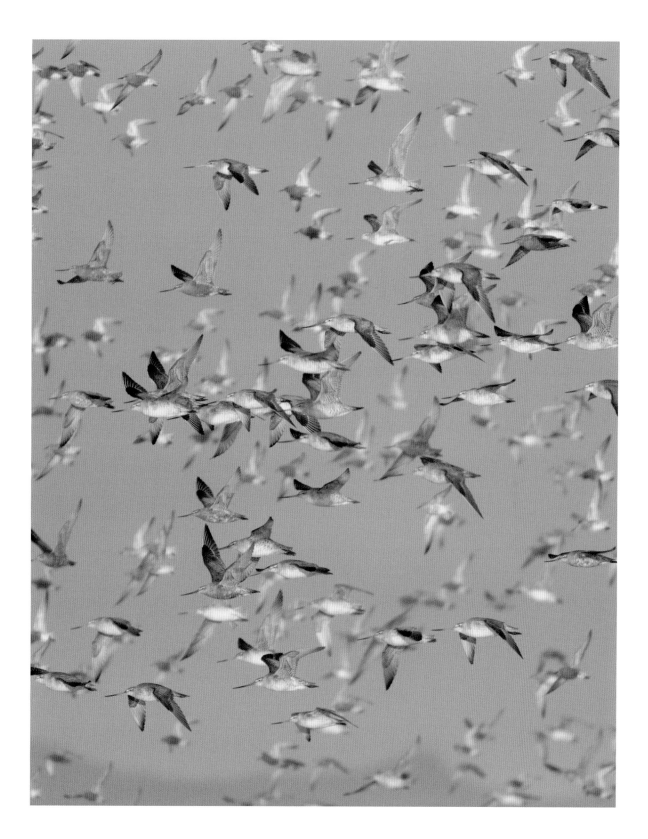

Fernbird

Shy swamp-dweller with tattered tail

Māori names	Latin name	New Zealand status	Conservation status
Mātātā, koroātito	*Bowdleria punctata*	Endemic	Declining

Scampering and scurrying through the dense undergrowth in wetlands, ever searching for bugs in the litter and scrub, the chestnut-and-brown fernbird looks for all the world like a foraging rodent. It even has a slightly gamey smell. Because of this bush-bashing life, its beautiful tail (which is as long as its body) is often worn down into tattered spikes.

If pressed, the fernbird will fly, but it'll stay low to the ground, short stubby wings whirring at full speed and shabby tail trailing awkwardly.

The fernbird is notoriously hard to spot, with clever camouflage and a shy nature making it easier to be heard than seen. It's highly inquisitive, though, and you might attract it for at least a second by hitting two small stones together or mimicking its high-pitched *tweet*. Some fernbirds are brave: the Snares Island subspecies will hop all over snoozing sea lions to nab the blowflies that settle around them.

Fernbirds may have once been one of New Zealand's most common birds (according to nineteenth-century ornithologist Walter Buller), but following widespread drainage of wetlands and the introduction of predators, such as mustelids and mice, they're in decline. They're now found in patches around the country in swamps, salt marshes, tussock flats, and various shrub or bracken

The Snares Island fernbird is often seen on the rocky shores of the main Snares island.

thickets. There are five subspecies named according to their island ranges — North, South, Stewart, Codfish and Snares. There was once also a Chatham Island species, which became extinct around 1900.

Early colonists called it swamp sparrow, but to early Māori the fernbird or mātātā was much more — he manu tapu, a sacred bird that served as a link between tohunga (priests or shamans) and the gods. Fernbirds could be good or bad omens depending on the activity, and they were used in a range of ceremonies. For example, at the burial of a deceased chief, men would catch a mātātā and perform a ceremony to lead the chief's spirit back to the ancestral home of Hawaiki.

Fernbirds often call in duet with their partner, or make a chittering or clapping sound with their beaks while quivering their feathers, all of which may be related to defending their territory borders. The usual territorial call is just one note, which is repeated and repeated for up to 15 minutes.

Fernbirds keep their territories all year round. They'll let a juvenile pass through, but only if it stays silent — any call and they'll chase it out. These juveniles will have to wait in nearby refuges with others until a vacancy for a territory comes up. At breeding time, both partners make a small, cup-shaped nest amid vegetation; less than a metre off the ground, it is woven from reed stems and fine grass and often lined with the feathers of other birds — anything from pūkeko to harrier, petrel or kākāriki, depending on who's nearby.

More often heard than seen, fernbirds usually stick to the insides of bushes and scrub rather than sitting out in the open.

Great albatrosses
Colossal wind-riders

Northern royal albatross

Māori name	Latin name	New Zealand status	Conservation status
Toroa	*Diomedea sanfordi*	Endemic	Naturally uncommon

Southern royal albatross

Māori name	Latin name	New Zealand status	Conservation status
Toroa	*Diomedea epomophora*	Endemic	Naturally uncommon

Antipodean albatross

Māori name	Latin name	New Zealand status	Conservation status
Toroa	*Diomedea antipodensis*	Endemic	Nationally critical

The largest seabirds in the world, the great albatrosses spend most of their lives soaring thousands upon thousands of miles around the Southern Ocean. These giants come in to land only once every couple of years to raise a chick, somehow finding again the small speck of ground where they themselves hatched.

The great albatrosses or toroa have inspired awe across the world's cultures. The first factor is their sheer size: they are among the largest of all flying birds. The southern royal albatross weighs about 10 kg and its wings span up to 3 m.

The second factor, linked directly to the monstrous size of their wings, is their ability to fly hundreds of kilometres with barely a flap of the wing. This lets them spend about 85 per cent of their life at sea — and for the first three years (at least) they don't touch land at all. They glide incredibly, moving about 22 m forwards while only dropping a metre.

There are no thermals in the cool latitudes of the Southern Ocean, so to travel great distances with very little flapping, they move in zigzag flight combining two techniques: wind-shear soaring (or dynamic soaring) and wave-slope soaring (a bit like using the updraughts on a ridgeline). In wind-shear soaring, the airspeed of an albatross gives it the power to stay aloft — as long as it keeps dipping, swooping and turning. The bird climbs into the wind, which is stronger at higher altitudes. It then banks

Antipodean albatross numbers have dramatically declined over the past decades, with many albatrosses dying as bycatch on Southern Ocean fishing vessels.

A special shoulder tendon 'locks' the wings into the outstretched position.

away downwind, losing height — then turns back into the wind close to the sea surface, where the wind strength is light enough (the boundary layer) that the bird keeps its forward momentum, and repeats the cycle. In this way, the different wind layers can propel the albatross on and on almost indefinitely, allowing it to spend its life at sea without using any energy on flapping. Albatrosses don't even expend energy holding their wings out: a special shoulder tendon 'locks' the wings into the outstretched position. This amazing ability to 'fly for free' using the wind holds potential in fields such as drone technology. Albatrosses do flap their wings in calm conditions, but it's so laboured

that they spend a lot of time sitting on the water on calm days.

Albatrosses have the lowest bird divorce rate in the world — usually only the death of a partner can break this enduring romantic bond. After up to two years apart, the male arrives first to prepare the nest, making a mound of greens and feathers bound together with mud. The female arrives too and soon she lays a single egg. Once the hatched chick has grown enough down to keep itself warm, and is large enough to fend off predators, it is left alone on the nest for monotonous months on end, seeing its parents only occasionally when they come back with food. After about eight months of sitting in solitude,

the youngster launches itself from the nest and flies out to sea — and stays there for anything from three to eight years, never touching firm ground. Albatrosses have salt-excreting nasal glands to allow them to drink seawater along the way; to Māori, the salt discharge seen dripping down the beak was ngā roimata toroa — 'the tears of the albatross', apparently weeping for its homeland (breeding ground) as it roamed far and wide.

The bird eventually homes back to its natal colony when it is ready to mate. It gathers on ridgelines with others of the same age, who are also on the hunt for a life partner. They stand opposite each other and the males do a courtship dance, screaming, tossing their heads and raising their wings. After testing their dance with lots of partners (which can take years), they eventually start spending more time with 'the one', and will begin to breed.

Understanding colony size is quite tricky. Albatrosses breed once every two years — so at any one time a whole cohort of breeding pairs may be out at sea having their 'year off' (but if their eggs fail one year because of a bad weather event, they'll try again the next). Also, all the young birds are out at sea for the first few years of life. This means numbers of great albatrosses present at a colony could be as small as one-tenth of the total population using that site.

LEFT
'Gamming' displays in large groups allow young southern royal albatrosses to practise their mating rituals and find a partner.

RIGHT
Kaikōura is the best place in New Zealand to encounter albatrosses, with daily boat trips to visit these wanderers in their ocean home.

There are six species of great albatross worldwide, and New Zealand has three of them, all of which breed only here: the northern and southern royal albatrosses, and the antipodean albatross. The southern is the larger of the two royal species, and both are among the longest-living birds in the world, with many making it to their forties.

Almost all northern royals nest in dense colonies in the Chatham Islands group — on the Big and Little Sisters and Forty-Fours islands. Northern royals became endangered after a vicious storm in 1985 stripped soil and plants from the Chathams nest sites — this left the birds with no nesting materials, so they laid straight onto rock: the perfect formula for a broken egg. It took more than a decade for the island soil and vegetation to recover enough for breeding to return to normal. However, the population there is still in decline because of this disaster: they take so long to mature, and breed so slowly, that the effects of 10–15 years of poor breeding can remain in the population for up to 40 years.

About 1 per cent of northern royal

albatrosses breed on the mainland, at Taiaroa Head on the Otago Harbour. The first egg was noticed there in 1920, but, despite further adult albatrosses arriving to nest, no eggs or chicks survived: egg-pilfering humans and hungry predators were to blame. But in 1937 a schoolteacher, Lance Richdale, set up camp and looked after the nests over the summer, protecting them from danger, and was rewarded with a chick that fledged. This was the first step in the eventual Royal Albatross Centre, which now has around 200 breeding royal albatrosses (along with little blue penguins, shags, fur seals, sea lions and a huge colony of red-billed gulls). One albatross at Taiaroa Head, named Grandma, raised her final chick at 62 years of age. The colony's chicks are, however, prone to fly-strike (blowflies lay eggs on the chicks' mouths and bums) and suffer from heat stress, and so the use of sprinklers and fly-repellents is essential.

Almost all southern royals breed spread out among the tussocks and herbfields of subantarctic Campbell Island, with a small number slightly further north on the Auckland Islands, and some even

Nesting among megaherbs (flora unique to the subantarctic islands), southern royal albatrosses prepare for a year-long investment in raising their chick.

flying north to Taiaroa Head where they hybridise with the northern royals.

Toroa have special value in the culture of Māori, who traditionally caught the birds out at sea; they ate the meat, made hooks and tips for spears from bones and used the feathers to decorate cloaks and canoes. The feathers are taonga (treasure) and were once worn, along with bones, by high-ranking people and for ceremonies. The wearing of the white feathers by Taranaki iwi symbolised unity and peace — and the concept that all the world's peoples were one — which culminated in Parihaka's non-violent resistance movement in the late nineteenth century. The Moriori of the Chatham Islands also wore toroa feathers as a symbol of peace and non-violence, which was the law of their historic chief Nunuku Whenua. Moriori traditionally harvested toroa eggs and chicks for food, from rocky islets around the islands.

The great albatrosses had meaning for Europeans too. Early explorers of the Southern Ocean were filled with awe by their immense wingspans and bodies, and the sheer distances they would cover, riding fierce winds with apparent ease. Sailors associated them with the dangerous gale-force winds in the latitudes of the Roaring Forties and Furious Fifties. Some endowed them with supernatural powers, believing that albatrosses were the souls of drowned seamen, and that killing one would surely bring bad luck. In Samuel Taylor Coleridge's poem 'The Rime of the Ancient Mariner' (1798), an old sailor shoots an albatross that had brought the ship favourable wind, and promptly runs out of wind and fresh water. Slimy creatures appear, his crew dies and other terrifying events take place.

The major threat to albatrosses is becoming bycatch on a fishing boat, when they scavenge for bait and fish. Some are more susceptible than others: for example, male and female antipodean albatrosses forage in different parts of the ocean to avoid competition in the non-breeding season. Female antipodean albatrosses are caught more often, resulting in a sex-biased population with fewer pairs able to breed. The other big threat, which is growing more prominent day by day, is plastic pollution: sadly, plastic-filled albatross corpses are an all-too-common finding.

Grey warbler

Tiny, long-suffering foster parent with the sweetest song

Māori names	Latin name	New Zealand status	Conservation status
Riroriro, kōriroriro, hōrirerire	*Gerygone igata*	Endemic	Not threatened

The grey warbler or riroriro is the little Pollyanna of the bird world, or maybe just one that lives in blissful ignorance of the fact it's being taken for a ride. For eternity it has been chosen by the shining cuckoo — a bird many times bigger than itself — to bring up an enormous alien chick, which kills the grey warbler's chicks in the process.

Yet the grey warbler warbles on like a happy cricket — gladly raising this greedy and enormous foster chick, and even hanging around after it fledges to make sure it's okay. And the warbler still keeps its own numbers extremely healthy even in today's fragmented ecosystem. It tops the charts as New Zealand's most widely distributed endemic bird; it prefers forest fringes but can be found in most places — except maybe alpine tussock. When they're not breeding, grey warblers can be seen in flocks (of usually five or six birds but occasionally more), sometimes with other insect-eaters, such as silvereyes.

The grey warbler male really does warble, and is extremely loud, given his tiny size (at 6.5 g, the grey warbler is New Zealand's second-smallest bird, after the rifleman). This warble is especially common in the breeding season when pairs are fiercely defending the territory they've peacefully held year round. (While the female will stick to her chirping contact call rather

The grey warbler male really does warble, and is extremely loud, given his tiny size.

than warble, she'll join in with the male on wild chases to get rid of intruders.) The song rings out clearly through dense foliage, but the bird itself is hard to see: it's tiny and grey, with olive-grey parts above and buff beneath, and a pointed black beak it uses to snap at prey.

The warbler is constantly active, jumping quickly from branch to branch hunting spiders, caterpillars, insect eggs and flies — sometimes taking small fruit, too. Sometimes it hovers vertically around the outer leaves of trees for four or five seconds at a time, wings beating extremely fast — something that no other New Zealand bird does. Adult grey warblers will stay in the same area for life; the only birds that move around are the juveniles, who'll move 3 or 4 km to find a new territory.

One possible reason this bird does well in New Zealand — despite 200 years of modified environment and introduced predators, and the much more ancient brood parasitism of the shining cuckoo — could be its unique nest structure, which is very hard for rats and other predators to penetrate. The female is responsible for this masterpiece. From about one year of age she can breed, and over about 27 days she carefully weaves a long, messy but spectacular blob; a pear-shaped creation suspended from the tip of a slender branch and sometimes also tied to others, often in trees or shrubs like kānuka, mānuka or gorse. Building materials include rootlets, moss, lichen, leaves, spider webs, bark, wool and whatever else she can find. Inside, the nest is lined with feathers and

Adult grey warblers have striking red eyes, distinguishing them from juveniles, who have duller brown ones.

More often heard than seen in the forest, grey warblers are our second smallest bird after the rifleman.

fluffy seeds, and to top it all off, the 3 cm entrance hole often has a little awning constructed over it to keep the rain off. The perfect home.

Once the female has finished construction and laid her eggs, she incubates alone. It's when the eggs are hatched that the male pitches in — he helps feed the chicks, especially if the female is moving on to organise her second clutch, which involves building a whole new nest-blob.

Early Māori used the nest of the riroriro to predict the coming weather. Amongst the Tūhoe people of the central North Island, if the nest entrance faced north, the spring's prevailing wind would come from the south, and crops would be plentiful. If it faced west, a warm east wind was predicted with mild conditions, and an east-facing entrance would bring cold westerly showers. For Ngāi Tahu of the South Island, if the nest was built high in the treetops, westerlies were expected; and if low, cold southerlies would dominate, which would be bad for crops.

For many, the grey warbler's song signalled that it was time to plant crops; there's a saying about people who are too lazy to plant seeds but try to eat the harvested crops:

I whea koe i te tangihanga o te riroriro, ka mahi kai māu?

Where were you when the riroriro was singing, that you didn't work to get yourself food?

A passage from Walter Buller's *History of the Birds of New Zealand* (1888) surely sums up the sunny nature of the grey warbler:

In the warm sunlight of advancing summer, when the manuka-scrub is covered with its snow-white bloom and the air is laden with the fragrance of forest flowers, amidst the hum of happy insect-life, a soft trill of peculiar sweetness — like the chirping of a merry cricket — falls upon the ear, and presently a tiny bird appears for an instant on the topmost twigs of some low bush, hovers for a few moments, like a moth before a flower, or turns a somersault in the air, and then drops out of sight again. This is the Grey Warbler, the well-known Riroriro of Maori history and song.

Gulls

Quintessential coastal wheelers

Southern black-backed gull

Māori names	Latin name	New Zealand status	Conservation status
Karoro, rāpunga	*Larus dominicanus*	Native	Not threatened

Black-billed gull

Māori names	Latin name	New Zealand status	Conservation status
Tarāpuka, tarāpunga	*Larus bulleri*	Endemic	Nationally critical

Red-billed gull

Māori names	Latin name	New Zealand status	Conservation status
Tarāpunga, akiaki, katatē	*Larus novaehollandiae*	Native	Declining

The shrill cry of seagulls is synonymous with the seaside: but they're not just squawking on the coast. Our threatened black-billed gull, for instance, will nest on remote braided riverbeds far inland, and the southern black-backed gull can be found ripping into rubbish bags at the local tip. Three gulls breed in New Zealand: the southern black-backed, the black-billed and the red-billed.

Southern black-backed gull

The southern black-back's scientific name, *Larus dominicanus*, refers to its pristine black-and-white garb, a bit like that of a Dominican friar. With a wingspan topping a metre, this large gull glides and soars elegantly on updraughts.

This gull is found all around the southern hemisphere — where it is often known as the kelp gull — on coasts, farms and ports, and even on lakes and rivers in the subalpine zone. It's also found in Antarctica. Overall its numbers are super-abundant, thanks to its scavenging skills: it homes in on anywhere scraps are available. It'll clean up the carcasses of dead animals at sea and on farms (where it eats afterbirth and lambs' tails, and will attack weak lambs, too), but it will also visit uncovered rubbish tips, fish processing plants, freezing works and kerbside rubbish bags, and this opportunistic streak has boosted its numbers to unnaturally high levels in some areas.

Black-backed gulls are also predators:

The largest gull in Aotearoa New Zealand is found all around the southern hemisphere, from the inner city to the Antarctic ice.

they'll eat fish and invertebrates, such as worms, but also small mammals and birds, eggs and chicks. They'll dig for shellfish, such as toheroa, dropping the big ones from up high onto rocks or roads to open them. They'll also eat insects inland, including crickets from pasture, or even perch high in rimu trees to eat the berries, a bit like a kererū.

They'll even attack whales. Off Argentina's Valdes Peninsula, these gulls have been recorded landing on the backs of right whales and gouging out chunks of skin and blubber, especially from mothers and calves. This has become more common over the last three or four decades (in the 1970s, 2 per cent of mothers and calves had these lesions; by 2011 that number stood at 99 per cent).

The Māori saying 'kai karoro' — 'seagulls' food' — was a morbid term for the defeat of a tribe in battle: where many were killed, the black-backed gull would feed on the remains of the dead. Early Māori also tamed these gulls, and sometimes used them to control plagues of kūmara moth caterpillars — they encouraged the birds to feed on the pests, which would otherwise decimate crops.

Along with the spur-winged plover, the black-backed gull is the only native bird not protected under the Wildlife Act in New Zealand. They were partially protected under the Wildlife Act of 1953, but this was revoked in 1970 because numbers had increased so much they were

If a chick wanders within the colony (or a fledgling practises flying away from the nest), it can be pecked on its head, killed and eaten by adults.

becoming a pest: they eat the chicks and eggs of threatened bird species, such as the black-fronted tern, black-billed gull, wrybill and banded dotterel, and they are also a danger to aircraft where colonies are close to airports. Periodically they are culled, using toxic bread baits or by the pricking of their eggs with formaldehyde.

The only places where black-backed gull numbers are still at natural levels are on remote island groups, where they nest in solitary pairs on headlands and rock stacks. Around mainland New Zealand, however, there are hundreds of colonies — on cliffs, offshore islands or shingle riverbed islands, some with up to a thousand members. They also breed in pairs on the tops of city buildings.

Adults usually pair with the same partner every year, beginning at the age of four or so. The male and female share the incubation and care of their chicks, which occupy a bulky nest of grass or seaweed, or a patch of bare sand. Each parent has a bright spot near the tip of its bill, which is what the chicks aim for when begging for food. In large colonies, things can get tense in the breeding season. While nesting birds will tolerate birds in closely neighbouring nests, if an unfamiliar bird steps too close, it will be attacked, struck with the bill and beaten with the wings. If a chick wanders within the colony (or a fledgling practises flying away from the nest), it can be pecked on its head, killed and eaten by adults.

Those that survive become mottled brown juveniles that are often mistaken for skuas. They moult into their adult colours at around three years old. These young have their own name in Māori — ngoiro.

The most delicate of our native gulls, black-billed gulls have a more restricted range than the other two species.

While more gracile than red-billed gulls, black-bills are no less shrill in their calls.

Black-billed gull

With a long, thin black beak, this is the most threatened gull in the world, and is found only in New Zealand. Unlike the other two New Zealand species, you won't often find black-billed gulls scavenging in city rubbish bins, as they tend to prefer natural food sources: when breeding in their noisy riverbed colonies, they eat invertebrates from rivers and pasture, taking flying insects on the wing and whitebait from the waterways. When wintering on the coasts, they take fish and marine invertebrates.

Why are they threatened? Their eggs and chicks are eaten by introduced predators — who also eat the adult gulls, given the chance — and by native predators such as southern black-backed gulls and swamp harriers. Black-billed gulls are also threatened by changes humans have wrought on their habitat. Most colonies are found on braided riverbeds in the South Island; where irrigation and hydroelectric dams have lowered the water level, weeds flourish on the riverbed, taking up nesting room and providing cover for predators. Colony sites are more of a mixed bag in the North Island, where the gulls nest on sandspits, at lakeside marinas and even around busy ports.

Red-billed gull

This is the most common gull on New Zealand coasts, and its brilliant red beak and legs are appropriate for its massive attitude. Like the black-backed gull, it likes

Red-billed
gulls live
in dense
colonies,
where
squabbles are
common.

to steal fish and chips and hang out at rubbish dumps. Its young can be mistaken for the black-billed gull, but the red-billed bird is never as elegant, nor is its bill so slender. Dense colonies are on the eastern coasts of New Zealand — on cliffs, craggy rock stacks and shores. The gulls are also found on subantarctic islands, breeding in smaller groups or singly.

Despite being the most common gull, it has declined in its three main breeding colonies (Kaikōura and the Mokohinau and Three Kings islands) since the 1960s, and in 2008 it was declared nationally vulnerable. Today, fewer than 30,000 breeding pairs remain. Introduced predators on the mainland are a factor, but when levels of krill (a major food source in the breeding season) decline, the gulls

also lose out. The one area where they are increasing is Dunedin, due to predator control around the colonies and more krill.

While they're normally coastal birds, there is a colony at Lake Rotorua; it is made up of both red-billed and black-billed gulls, which do interbreed sometimes — and create fertile hybrids. This colony is important to some Māori: the story goes that the Arawa tribe thought themselves safe from attack on Mokoia Island, but Ngāpuhi sought to surprise them by dragging their canoes inland to Rotorua to get to the island. Their plan was foiled, however, when the gulls spied them and gave the alarm (although even still, many Arawa people were killed). The gulls then became tapu to Te Arawa and were protected from harm.

Kākā

Talkative tree-shredder

Māori name	Latin name	New Zealand status	Conservation status
Kākā	*Nestor meridionalis*	Endemic	Recovering

The kākā is a forest parrot, large and olive-brown, with a crimson tummy and bright red or orange under its wings. It's not quite as naughty as fellow parrot the kea, nor as heavy as the kākāpō, but this loudmouth gives them a run for their money in the personality stakes.

Its screams, chatters, and calls of *ka-aa* (which can sound like fingernails on a blackboard) echo over forest canopies as the kākā flaps around in raucous flocks, punctuated unexpectedly by almost liquid whistles that are sweet and musical. It's such a noisy bird that a Māori term for someone ditzy and talkative is he kākā waha nui, 'a big-mouthed kākā'.

Chatter aside, this bird has a weapon of a beak, using it to chisel greedily into trees to extract insects such as huhu grubs and other beetles. The kākā also loves sap, which it laps up like an elixir, tearing pieces of bark off trees to get at it, or wielding its powerful beak like a can opener to make horizontal gouges in the bark and lick what oozes out. For thick bark, sometimes the kākā uses its entire body as leverage and expends considerable time and energy to get at the good stuff. Unfortunately this behaviour can kill sensitive trees. Wellington's predator-free sanctuary Zealandia reintroduced kākā to the area from 2002 (after they were extinct in Wellington for over 100 years), and since then they've grown to number in their hundreds. Despite general support for these rambunctious birds from Wellington residents, the damage is being felt by the city's conifers and eucalypts, including culturally important trees in the

Enjoying
the rain in
Wellington.

Chatter aside, this bird has a weapon of a beak, using it to chisel greedily into trees to extract insects.

Botanic Garden. Kākā are loving the exotic trees, but in remote native forests the scars on trunks can be seen also in māhoe, kānuka, rātā and pōhutukawa, and on conifers like rimu and tōtara.

Like the tūī, the kākā has a brush-tipped tongue, great for that sap slurping but also for sipping nectar from flowers such as kōwhai, rātā and flax. In beech forests, kākā visit tree trunks to lick honeydew secreted from scale insects; the energy from this sugary substance is suspected to be important for kākā breeding, but unfortunately the invasive German wasp eats a lot of the honeydew, too. Kākā also love berries of the native trees hīnau, miro and tawa, and in Wellington city they eat residents' fruit and nut crops. Like the kea, they have a destructive streak, occasionally damaging joinery, cladding and chimneys.

Kākā, after the kererū, were one of the birds most sought after by Māori for food — especially when they were gorging on the nectar of rātā in full bloom. Hunters speared them or caught them in a noose. They'd often use a live decoy — a screeching pet kākā, tied via a leg ring to a perch — to lure curious wild birds in. They preserved kākā in their own fat, and used the scarlet feathers from under the wing for feather cloaks. These feathers were much in demand and featured in folklore — according to one Māori legend, the kākā hides its red feathers under its wings

LEFT
From sibilant whistles to raucous chatter, kākā fill their forest homes with constant commentary.

RIGHT
Kākā on the South Island are larger than those in the north, although there is no genetic basis for subspecies within the population.

because it stole them from the kākāriki (after the kākāriki itself had brought them all the way from a mythical ancestral land). Historically, there was the odd kākā that was all red or white — the white ones in particular made great decoys to attract other kākā.

Kākā can fly fair distances. For example, all of the kākā on the Hen and Chickens, Little Barrier/Hauturu and Great Barrier/Aotea islands belong to the same population, though they are spread around the Hauraki Gulf. Kākā used to absolutely cover the mainland, swooping in clamorous flocks of hundreds of birds. By the late nineteenth century numbers had plummeted — helped along by forest clearance and widespread hunting by

both Māori and Pākehā (early European miners and settlers valued it as food, like the Māori) — and the kākā has been fully protected since 1907.

As with many New Zealand native birds, the kākā's nesting style is its downfall in a land now filled with introduced predators. The kākā favours hollows in the branches and trunks of trees, hacking into the timber to make a cosy nest of woodchips, which, unfortunately, makes it easy for stoats, rats and possums to sample the eggs. Stoats will also kill chicks and even the incubating female kākā. The biggest kākā populations are now all on islands without predators, or mainland sites where predator control is carried out.

Kākāpō

Fragrant forest bumbler

Māori names	Latin name	New Zealand status	Conservation status
Kākāpō, tarapo, tarepo	*Strigops habroptilus*	Endemic	Nationally critical

The kākāpō breaks a few records. It's the world's heaviest parrot (males can reach 4 kg), and the only flightless, nocturnal parrot (its name means 'parrot of the night'). It has the slowest metabolic rate of any bird and may live for almost a century.

Bumbling around the forest, the kākāpō uses its extremely strong toes — two forward-pointing and two pointing backwards — to climb some 20 m up into rimu trees to feed and sometimes sleep. But its wings are much too short for its size to fly down, so it plummets to the forest floor, flapping its wings to make the fall slightly more controlled than nothing at all. The kākāpō can, however, hike solo for kilometres at a time on its strapping legs, which is unusual for a parrot.

Kākāpō are critically endangered. In 2019, there are only about 150 adult birds in existence; this is at least up from the 1990s, when they numbered around 50.

However, the 2018–19 breeding season was an exceptionally good one, with 86 eggs hatching and 77 chicks alive at the time of writing. While some of these chicks won't make it, a high proportion is expected to survive to adulthood — an incredible increase to the kākāpō population. How did this happen? The kākāpō is vegetarian, eating leaves, seeds, roots and forest fruits from trees such as rimu. Rimu fruit are linked tightly to kākāpō breeding — the birds nest only in the 'masting' years, every two or three years, when the rimu trees bear a lot of fruit. Something in the unripe fruit — possibly the high levels of vitamin D and calcium, which are both

All kākāpō have names, and carry transmitter harnesses that allow researchers and conservationists to monitor them.

Because each egg is so critical to the species, if any are cracked, kākāpō workers will tape them up.

important for egg and chick development — triggers the females to nest. Then, the bumper crop of rimu berries in the summer means the chicks have plenty to eat. The 2018-19 masting season was a particularly good one.

Before humans arrived, kākāpō numbered in their hundreds of thousands, dispersed around the whole of New Zealand. Being flightless and prone to 'freezing' when startled worked well when the camouflaged kākāpō was hunted from above by its natural predators, such as harriers, the falcon and the now-extinct Haast's eagle. But when introduced predators began to hunt it on land, its fate was sealed. Moreover, the kākāpō has a unique, musty-sweet smell, which even humans can detect (it's reminiscent of peach), so for keen-nosed killers like stoats this odorous bird was easy meat.

Kākāpō were eaten by both Pākehā and Māori. In 1899, explorer Charlie Douglas said they could be caught at night on low scrub 'by simply shaking the tree or bush till they tumbled on the ground, something like shaking down apples. I have seen as many as a half a dozen Kakapos knocked off one tutu bush this way.' Douglas himself ate them, boiling the old tough ones or putting them in a Māori oven. Māori would hunt them with dogs, or set spring-snares, or simply sit beside the tracks that kākāpō make in the bush at night — which were as obvious as sheep tracks — and use light to startle the birds into freezing.

Māori would only occasionally use their feathers for cloaks: when someone complained too much of the cold, others might use the proverb, 'Me kauhi rānei koe ki te huruhuru kākāpō?' (Shall I cover you up with a cloak made of kākāpō feathers, heaped up here from the south?), which implied, 'Would nothing else be good enough for you?'

Hunting saw the kākāpō die out in the North Island by 1930. Meanwhile, the Europeans with their weasels, stoats and cats had all but killed them off in the south, too. In the 1890s, conservationist Richard Henry had spent years translocating hundreds of birds to Resolution Island in Fiordland, in New Zealand's first island translocation operation, only for stoats to swim out to the island and undo all his hard work.

Intensive conservation programmes began in the 1980s–90s, moving birds to more secure islands. In the breeding season, volunteers and scientists would camp near nests, using tiny electric blankets to warm up unattended nest eggs and chicks, and letting off elaborate smoke-and-sound firecrackers to deter any predators seen on camera. They've stopped with the heat pads now — when a mother leaves, her chicks go into semi-torpor and become unnervingly cold, but researchers have since discovered this is a normal adaptation and they soon warm up again when she returns. Because each egg is so critical to the species, if any are cracked, kākāpō workers will tape them up. In 2014, on Codfish Island/Whenua Hou, an egg was so crushed it was doubtful it would survive, but tape and glue worked a charm and the resulting chick, Ruapuke, survived and thrived.

The kākāpō's mating habits are strange. Unseen in other parrots, kākāpō have a lek mating system — lek means 'play' in Swedish. When it's time to mate, they stop being hermits and the males congregate within hearing distance of each other. (But they don't get too close: if two males meet, they may fight to the death.) Each male digs out a bowl on the ground, with a carefully cleared track leading to it. He sits in his bowl, puffs himself right up, inflates a thoracic air sac, and starts 'booming' every few seconds, for up to a minute before starting another sequence — and can keep this up for eight hours. The booming is an un-bird-like, pulsating low-frequency sound that can be heard up to 5 km away. He then pauses to listen out for females. He will also make a higher-pitched 'chinging' sound, perhaps so that females can locate him more easily. The males do this every night for up to three months. Females will come for a brief fling, most mating with the same male, and his work is done — he'll shuffle off to be a hermit again, leaving the females to incubate and raise the chicks alone.

Weight matters a lot for a kākāpō mother — it can dictate whether she has a boy or a girl. When she's well fed before laying, she's more likely to produce males. A

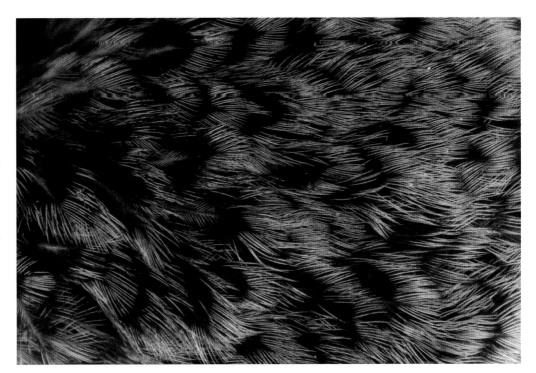

Kākāpō are extremely well camouflaged in their forest habitat, with plumage patterns reminiscent of moss.

possible reason for this is that the male chick will also be well fed and grow up strong and able to outcompete other males, enabling him to mate with many females and father more offspring. But if she's under-nourished, female chicks are more likely.

Kākāpō are generally so fearless around humans — they enjoy climbing up our arms and even preening our heads — that the New Zealand govenor George Edward Grey said the birds' playful affection was 'more like that of a dog than a bird' and that it would make a great pet if not for its 'dirty habits'. Kākāpō were catapulted to international fame when, in 2009, a bird named Sirocco got frisky during the filming of a BBC television series with Stephen Fry. Sirocco climbed up onto zoologist Mark Carwardine and enthusiastically mated with the back of his head while clapping the man's face with his wings, prompting Fry to remark, 'You are being shagged by a rare parrot.' Sirocco became a celebrity, and in 2010 Prime Minister John Key made him New Zealand's official spokesbird for conservation. Sirocco is ideally suited for public appearances — he was hand-reared as a chick away from other kakapo, leaving him imprinted on humans. He even 'booms' in the presence of people rather than with other kākāpō.

Kākāriki

Rare in the wild, but caged around the world

Red-crowned parakeet

Māori names	Latin name	New Zealand status	Conservation status
Kākāriki, kawariki, porete	*Cyanoramphus novaezelandiae*	Endemic	Relict

Yellow-crowned parakeet

Māori names	Latin name	New Zealand status	Conservation status
Kākāriki, pōwhaitere, torete	*Cyanoramphus auriceps*	Endemic	Not threatened

Kākāriki is the Māori word for both 'green' and 'little parrot', and one look at this dazzling emerald bird explains the linguistic connection. So how do you find a kākāriki in a green forest canopy? If you're in the forest, listen for an incredibly noisy munching and cracking, and follow it to a rain of debris falling from the treetops — you'll be covered in chewed bark, leaves, fruits and flower petals.

Alternatively, they may be down on the forest floor, scratching like a chicken in the leaf litter for insects, small pebbles and dirt. Or if you're on Antipodes Island, head to the pungent penguin colonies, where this tropical-looking bird can be found unexpectedly scavenging carcasses and broken eggs — or even digging up poor storm petrel chicks from burrows for a tasty snack, a bit like a kea. Or you could simply go online — the kākāriki has long been a pet, sold all around the world, the only New Zealand native bird in this situation!

The kākāriki is New Zealand's versatile parakeet, found from the subtropics to the subantarctic region. There are eight species found around the South Pacific, and New Zealand has six of them, from the Kermadecs all the way to the chilly

LEFT
Smaller
than their
red-crowned
cousins,
yellow-
crowned
kākāriki have
a slightly
higher-pitched
chatter.

RIGHT
Far from the
tropics, red-
and yellow-
crowned
kākāriki
inhabit the
Auckland
Islands, where
they readily
interbreed.

Auckland and Antipodes islands. These six are the yellow-crowned, orange-fronted and red-crowned kākāriki; Reischek's parakeet; Forbes' parakeet (like the yellow-crowned but found only on an island in the Chathams); and the Antipodes Island parakeet. The green feathers are fairly signature, but the different species have unique patches of red, yellow or orange, or flashes of blue.

Kākāriki have low numbers today, as a result of hunting, culling, burning of habitat, and predation by rats, cats and mustelids. Before humans arrived, their threats were limited to predatory birds such as the morepork, laughing owl (now extinct) and falcon, and fossil deposits show kākāriki used to exist in immense numbers. In the late nineteenth century, there were so many that if their forest food ran out they were considered a nuisance in rural areas, especially in the South Island. Plagues of them would arrive down from the forests and consume wheat and orchard fruits. They were shot in their thousands and their vibrant feathers were even used to stuff mattresses.

The plentiful kākāriki inspired many a Māori proverb. Their name was almost synonymous with 'numerous' — for instance, if someone was being bothered by lots of inquisitive children, they would compare them to those tiresome flocks of invading kākāriki. A gossip or noisy chatterer would be compared to a brood of crying kākāriki nestlings in a hollow tree.

Someone who ate as soon as they got up in the morning was called a kākāriki (Māori traditionally did not eat first thing). The kākāriki was also a food source for Māori, who would bait a trap with berries to lure the bird in, to be caught by snares. The hunter would then use these live birds as decoys to lure other curious kākāriki closer. Even today, kākāriki living on islands where they rarely see humans will be quite unafraid and may come right up close to investigate.

Kākāriki may not be so abundant now in the wild, but there's certainly no shortage of captive birds. Red-crowned and yellow-crowned kākāriki are the only New Zealand native birds that can be legally kept in captivity in New Zealand (with a permit from the Department of Conservation) by members of the public, because they have long been popular pets around the world. This began in the nineteenth century, when they were exported to Europe and other parts of the world, and in New Zealand too red-crowned parakeets were once a very common caged pet in households. A search online reveals countless videos of kākāriki antics in homes worldwide — and in Australia you can buy a young kākāriki for as little as $50. They've even been bred for different colours, including a deep sky blue. Today, New Zealand no longer permits the import or export of kākāriki. Nor may pet kākāriki be released into the wild in New Zealand, because they hybridise in captivity and could also transmit parrot diseases. (They can interbreed in the wild, too, but this rarely happens.)

It's thought the yellow-crowned kākāriki now has the highest numbers, but the red-crowned birds are found over a larger area. And certain sites buck the trend: on Raoul Island, for example, there are no yellow-crowned kākāriki but thousands of reds. The orange-fronted kākāriki was once found throughout New Zealand but it is now the rarest parakeet, with only 100–300 left in the wild, on predator-free islands and in a couple of beech forests in the South Island: introduced predators and habitat destruction are to blame. A captive breeding programme in Christchurch is helping to keep numbers afloat by releasing birds raised by foster parents.

The yellow-crowned kākāriki can be found in tall dense forests on the three main islands and a number of smaller offshore islands, whereas the red-crowned prefers open spaces near forests. Kākāriki are normally found in pairs, but a couple of months after the breeding season you'll see family groups of six or seven birds; after that, once they've been kicked out of their parents' territory, juveniles will flock together.

Kākāriki can nest high in the canopy, in the hollow limbs of big old trees like pūriri, taraire and kahikatea (unfortunately this is also where rats and possums like to 'nest' — especially if there are birds' eggs to eat, too). The birds also use holes in cliffs, in the ground or among vegetation. The Antipodes Island kākāriki, bereft of trees, will nest under tussock.

If food is plentiful, kākāriki can nest right through winter, but usually this takes place from October until December. The female does nearly everything: nest-building, incubation, and the brooding and rearing of chicks in the first couple of weeks. While she's busy, the male feeds her by regurgitation, until the chicks are 10–14 days old and he can help out with them instead. The seven or so chicks hatch at different times, so they range greatly in size. Often, red-crowned parakeets will feed fledglings on the ground before they can fly. One Māori legend says that once the chicks have hatched, green lizards called moko kākāriki spawn from the broken eggshells.

Red-crowned
kākāriki,
Chatham
Islands.

Kea

Alpine prankster with sky-high IQ

Māori names	Latin name	New Zealand status	Conservation status
Kea, keorangi	*Nestor notabilis*	Endemic	Nationally endangered

The world's only alpine parrot, this mischievous bird seeks out human contact for food and fun. It's found throughout more than 3.5 million hectares of the South Island from north to south, from lowland podocarp forest to alpine beech and mountain scree, but you'll hear its call of *keeeeeaaa* the most where human and kea habitats overlap — skifields, roads, huts and high-country farms.

The kea's a big bird of mostly olive green plumage, with fiery orange under its wings and rump, and a blue-green iridescence to its wings.

And it's a rascal. It will eat your shoes while you're wearing them, strip the rubber off car windscreen wipers and door seals, and tear insulation off power lines. It will empty rubbish bins, snip through tent guy-ropes and seams, and slide noisily down iron roofs for fun. It has even hung onto the rotors of a helicopter as the pilot slowly started them turning (after the birds were trying to pull out important wiring), eventually being joyfully flung off in the ultimate showground ride. In one instance of vandalism, a kea and a weka teamed up to raid the Wobbly Kea Cafe in Arthur's Pass at night. Security camera footage showed them breaking in through the cat

The world's only alpine parrot is at home in the harsh conditions of Aotearoa's mountain regions.

Dubbed 'the sheep-eating parrot', the kea was held responsible for hundreds of missing sheep.

door, the kea first and the weka following, after which they wreaked havoc in the kitchen — throwing implements around, pilfering food supplies and generally destroying property.

Destructive urges aside, the kea is considered to be one of the most intelligent birds in the world. It is able to cooperate at a level similar to chimps and elephants, as shown in an experiment where two kea could drag a board holding food into their enclosure if they both pulled the string at the same time. One would even wait for a minute until the other bird arrived to help.

Kea have such a predilection for discovering how things work, how they break and how they taste, that 'kea gyms' were set up near the Manapōuri Power Station (in an effort to stop them damaging cars in the car park) and next to roadworks in Fiordland, Nelson and

Arthur's Pass (to stop them moving roadworks cones around dangerously between traffic). It's thought that they might be moving cones not just for fun, but to slow cars down to beg for food. The kea gyms have ladders, spinning flotation devices, swings and climbing frames, which are rearranged regularly to keep the birds interested; however, it's hard to trick a kea: reports are the kea soon move on to other exciting pursuits.

Aside from *keeeeeaaa*, kea have a special shriek that puts them in playful mode no matter their age — much like how infectious laughter works with humans. When they hear the call, they'll start playing with other birds, playing chase and throwing things, or even just entertain themselves with aerobatics.

The kea eats more than 200 types of food. Its keen sense of smell is useful for

finding buried roots. It also eats bulbs, seeds, leaves, insects, mammals and other birds and eggs. But it loves fruit and, fittingly, New Zealand's alpine areas have an unusually high number of fruit-bearing plants. Twelve per cent of alpine plants rely on kea for seed dispersal.

Kea chicks face a cold start: the parents incubate and hatch their chicks in crevices under rocks and in holes between tree roots in winter or spring, when there's still snow on the ground (unlike their cousin the kākā, which breeds in spring-summer). Breeding pairs claim a whole spur of a hill as their territory. Meanwhile, juvenile males (you can tell young birds from the yellow eye ring, cere where the nostrils are, and lower mandible) range far in gangs,

wreaking havoc on anything interesting that they find, until they settle down at three or four years old. When their flock passes through the territories of adults, the adults will join the flock and be quite tolerant of the juveniles — the hierarchy is very flexible — which allows the younger ones to hang out with the seniors and pick up a few essential life skills, such as how to forage. The oldest known kea in captivity was 50 years old, and some in the wild have reached their late twenties.

While adaptability is the kea's main strength, it has brought the bird many enemies, too. Since humans arrived, it may have learnt to use its bill (unusually slender for a parrot) to dig for grass grubs in agricultural pasture, and huhu grubs

Many kea now carry alpha-numeric bands, so researchers can get an idea of the population size.

from rotten logs in pine forestry, but it also scavenges deer shot by hunters, makes messes at rubbish dumps, and begs for food at ski fields — stealing hats, gloves and cameras in the process.

In fact, there used to be a bounty on kea — up to one pound per beak ($120 today). Between 1860 and 1970, 150,000 kea were shot because some had developed a taste for mutton fat. Dubbed 'the sheep-eating parrot', the kea was held responsible for hundreds of missing sheep after a winter up in the high country — even though harsh conditions and poor animal welfare could also have been to blame. At first, when settlers in the late 1860s found their sheep dead with raw flesh on their back, they thought their sheep had a disease. But it was eventually discovered the 1 kg kea could hold onto the wool of the live sheep and drill into the blood and flesh, accessing the fat around the kidneys. The sheep would often die even from small wounds because of blood poisoning from bacteria in the parrot's beak. There's a theory the kea might have thrived on carrion from dead moa before the extinction of moa in the seventeenth century, and so when sheep turned up, providing a new source of fat and protein,

the kea's numbers also shot up. Sheep killing was usually done by just a few rogue kea — with others scavenging the carcasses — but they were all persecuted. Bounty hunting to protect sheep was banned in 1971, and full protection was given to kea in 1986. Today, it is the Department of Conservation, not the landowner, that deals with nuisance kea.

But kea are still declining, and they're now endangered. It's estimated their population has crashed 50–80 per cent since the early 1980s, and now their numbers could be anywhere between 1000 and 7000 — they're hard to count because they're unevenly distributed throughout the South Island high country, and they can fly long distances in flocks.

Introduced predators eating chicks and eggs are now the biggest killers of kea, but these parrots are also killed by cars, snow groomers and power lines, and by eating human food. Another cause of death or brain damage is lead poisoning — kea love chewing the sweet-tasting lead fixtures on old back-country huts and mines. DOC is working on removing lead from buildings, but it's an enormous job.

Kererū

Plump waistcoated pigeon with whooshing wingbeats

Māori names	Latin name	New Zealand status	Conservation status
Kererū, kūkū, kūkupa	*Hemiphaga novaeseelandiae*	Endemic	Not threatened

The kererū is no ordinary pigeon — it's magnificently clothed, very fat and it sometimes gets drunk on fermented fruit.

One of the largest pigeons in the world (two to three times heavier than the rock pigeon), the endemic kererū sails loudly through the air on whooshing wings, wearing a white waistcoat. It lands on flimsy branches that sink under its weight and proceeds to stuff its face with foliage, flowers or large fruit, eaten whole. The bird will often have so much to digest, it'll sit in the sun for a long time while its crop is hard at work — so long, sometimes, that the fruit begins to ferment and produce alcohol. Soon, completely over the limit, the plastered pigeon will simply fall out of the tree, or it'll attempt to fly, and crash into windows or drop into traffic. Luckily for all, drunk kererū are a rarity.

Gluttony and getting trashed aside, this glossy blue-green and purple-bronze beauty is vital to our New Zealand mixed podocarp–broadleaf forests. This is because, while it eats many types of fruit (at least 70 species), it's the kererū's ability to stretch its beak and throat around the biggest of fruits that really counts. It is now the only common New Zealand bird that can swallow a whole fruit more than 10 mm in diameter (picture us eating a grapefruit whole), then pass it intact and spread its seeds far afield. Without kererū, native trees like karaka, tawa, taraire, matai and pūriri could no longer colonise new locations and would instead just grow in clumps.

Making a precarious and draughty-looking nest of twigs, kererū raise only one chick at a time, but if food is plentiful they can occasionally raise up to three consecutively in one breeding season — even if it means having a chick in one nest while incubating an egg in another. Like other pigeons, the male and females feed 'crop milk' to the chick when it's tiny; this cottage-cheesy substance, which comes

In different lights, the plumage of kererū reveals an iridescent mix of purples, fuchsias, greens and blues.

French explorer Julien Crozet wrote admiringly in 1772 that 'in the forests are very beautiful wood pigeons about the size of a pullet, [and] their sparkling blue and gold plumage is magnificent'.

from the walls of their crops, is chocka with protein and mixed with fruit pulp.

Kererū were an important food source for Māori, who valued them as taonga — they were so big, plentiful and delicious, and their feathers were used to decorate weapons, waka, houses and cloaks. Preserved kererū, or huahua, were a delicacy and were given as gifts or traded. The bird's fat, mixed with mashed fern root, was made into cakes for travel. It was also mixed with speargrass gum to make a scented oil.

Early Māori had ingenious ways of taking them: one of these used the bird's thirst. Once kererū had gorged themselves on fruit, such as miro, they'd be parched and would head for a drink. Bird catchers would place rows of nooses alongside streams or water troughs to catch the birds by the neck. They'd also spear them, or set snares, or simply use a long pole to slip a noose over their neck as they ate. Māori fowlers were extremely good at getting very close to birds. All of these methods would be outlawed when kererū were added to the game list.

Europeans appreciated the kererū's potential for the pot. French explorer Julien Crozet wrote admiringly in 1772 that 'in the forests are very beautiful wood pigeons about the size of a pullet, [and] their sparkling blue and gold plumage is magnificent'. Walter Buller wrote of the

LEFT
Kererū are the only birds big enough to disperse the seeds of large-fruited native trees such as pūriri, tawa and karaka.

RIGHT
Kererū are impressive in flight, and perform stunning aerial displays during the breeding season.

scale of trapping: 'Its relative abundance may be inferred from the fact that in July and August 1882, Rawiri Kahia and his people snared no less than eight thousand of them in a single strip of miro bush about two miles in extent by half a mile in width at Opawa near Lake Taupo.' Māori would not, however, hunt kererū when they were feeding on kōwhai shoots, because apparently eating the meat would bring on terrible headaches.

People made a living shooting kererū and selling them to goldminers on the West Coast. Pigeon pie and jugged pigeon were popular. Kererū were also shot for sport by Pākehā, but since the birds are relatively unafraid of people, they didn't make for a very exciting game bird.

Kererū have been protected since 1912, but they still suffer from poaching by both Māori and Pākehā. The biggest threats today are stoats, possums, cats and rats, all of which will eat the eggs and young out of the nest; stoats and cats will even attack and kill adult kererū. Possums also devour most of the kererū's preferred fruits, shoots and leaves — both native and exotic, such as tree lucerne, broom, willow and kōwhai.

Despite these threats, kererū are common throughout the three main islands and on smaller islands, although numbers vary between regions. They're hard to count, because they're difficult to tell apart and have large home ranges, and travel almost 20 km from their home range for food. The species has been at the centre of a nationwide annual citizen science project, the Great Kererū Count, since 2011. For a week or so, people report sightings of kererū, and the data received provides a better idea of numbers in different parts of the country, as well as of kererū behaviour.

Some Māori today are keen to resume customary harvest of kererū, but so far most Māori and non-Māori agree it is unsustainable because numbers are too low. Some iwi are working with government and other agencies to increase kererū numbers in the hope that someday they might sustainably harvest them and keep the tikanga alive.

The Chatham Islands has a very similar species, known as the parea. It is about 20 per cent heavier than the kererū, and is found only on the main island.

Kingfisher

Pint-sized battering ram

Māori names	Latin name	New Zealand status	Conservation status
Kōtare, kōtaretare, kōtarepopo	*Todiramphus sanctus*	**Native**	**Not threatened**

The kingfisher is a coiled spring, perching hunched and motionless on power lines, gimlet eye scanning for any small movement of prey, huge beak dwarfing its small head. Bang! Suddenly it's a dazzling turquoise streak through the air, snatching a beetle from the ground. Quick as a flash it has returned to its perch, where it smashes the fight out of the hapless beetle before swallowing it whole. Later, it regurgitates the indigestible parts.

New Zealand's native kingfisher, kōtare, is a slightly larger subspecies (*vagans*) of the sacred kingfisher, which is found in Australia and parts of the western Pacific. Its technical name 'sacred kingfisher' is used because a very similar bird, the extinct kaua, was sacred to the god Tāne amongst the indigenous people of Rarotonga. The sacred kingfisher is just one of 86 species of kingfisher worldwide, which come in all colours and hues.

Kōtare have incredible eyesight. From their high perches — around 2 m or more off the ground, such as a washing line, a tree, a power line or a fence — they can spot small prey from a large distance. In Māori culture, the saying 'he kōtare koe!' describes someone hungrily watching a

Crabs are a favourite of estuarine kingfishers, who will beat them mercilessly into submission before swallowing them whole.

meal, waiting to be invited — just like a kingfisher hunched on a perch, watching for food. While kōtare sometimes snap up insects from tree foliage, their prey is usually on the ground or up to a metre under the water — fish, freshwater crayfish and tadpoles in freshwater, small crabs on the mudflats. Inland, they'll gobble up cicadas, beetles, wētā, spiders, lizards, and even small birds and mice. Ornithologist Geoff Moon described seeing a chick having just been fed a mouse by its parents: the mouse's tail was hanging out of the chick's beak for half an hour before the head was digested enough for the chick to swallow the rest.

Kek kek kek is its loud territorial call, uttered when anyone, including another kingfisher, is approaching its nest too close. Dive-bombing sometimes follows. Historic news reports paint these birds as having a vicious side. Writer Murdoch Riley saw news reports in the early 1900s of kingfishers attacking a cat that was in a tree near their nest, bayonetting it with their bills until it fell to the ground and later died from its wounds. He described another story of a kingfisher crushing the skull of a singing tūī, killing it, then flying away. Accordingly, kōtare are often

mobbed by small birds — and in the past they've come in for persecution from humans. In 1860s New Zealand, plagues of caterpillars were devouring pasture and in some cases even stopping some trains from gripping the tracks, because most New Zealand native birds preferred to stay in the bush rather than forage on pasture. To combat the caterpillars, sparrows were brought in. When acclimatisation societies introduced the first sparrows from Australia and Europe, at great cost, kingfishers would attack them — and so they put a bounty on kingfishers. Destroying them was encouraged, too, to protect newly introduced trout.

Kōtare are often spied in patches of forest near the coast, on farmland and in inland forest. In winter, when inland insect supplies dwindle, inland kōtare will fly to the coast to take advantage of marine food — they're often seen around estuaries swooping down to take mudcrabs without pausing to land.

In the spring-summer breeding season, they nest in rotten tree trunks or hollows, or riverbanks and cliffs — you may spot a small hole, its entrance streaked with guano, and hear the sandpaper scream of chicks from within. Although kōtare

can use the same cliff burrow or tree hole for years, sometimes they need to make a new one. They do this by perching on a branch a few metres away, then flying at high speed straight into a bank or small tree hole, smashing into it with their strong, chisel-like beaks until they make a dent. Once they can perch in it, they scoop and peck out the rest to make a tunnel. The tunnel slopes slightly upwards as a guard against rain, and they hollow out a chamber at the end of it. The female and male create the hole together, working in tandem for a few days to get the job done — longer if it's a tree to be chiselled. They start incubating as soon as the female lays her first egg, so the chicks hatch one after the other and are different sizes in the nest. After all this work, sometimes the family can be evicted by other birds, such as mynas.

Within traditional Māori culture, kōtare were used for feathers and food. Before it gained protection, the kingfisher's blue feathers were in high demand among Pākehā as hat ornaments and fishing lures. Early New Zealand ethnographer Elsdon Best wrote, however, that Māori did not use the blue feathers in traditional garments — preferring to use red, brown, black and white feathers. Some iwi would cook fledglings in a hāngī, but other tribes would not eat them because the kingfishers ate lizards, which were thought of as evil spirits.

A flash of
azure across
the mudflats,
Whangateau.

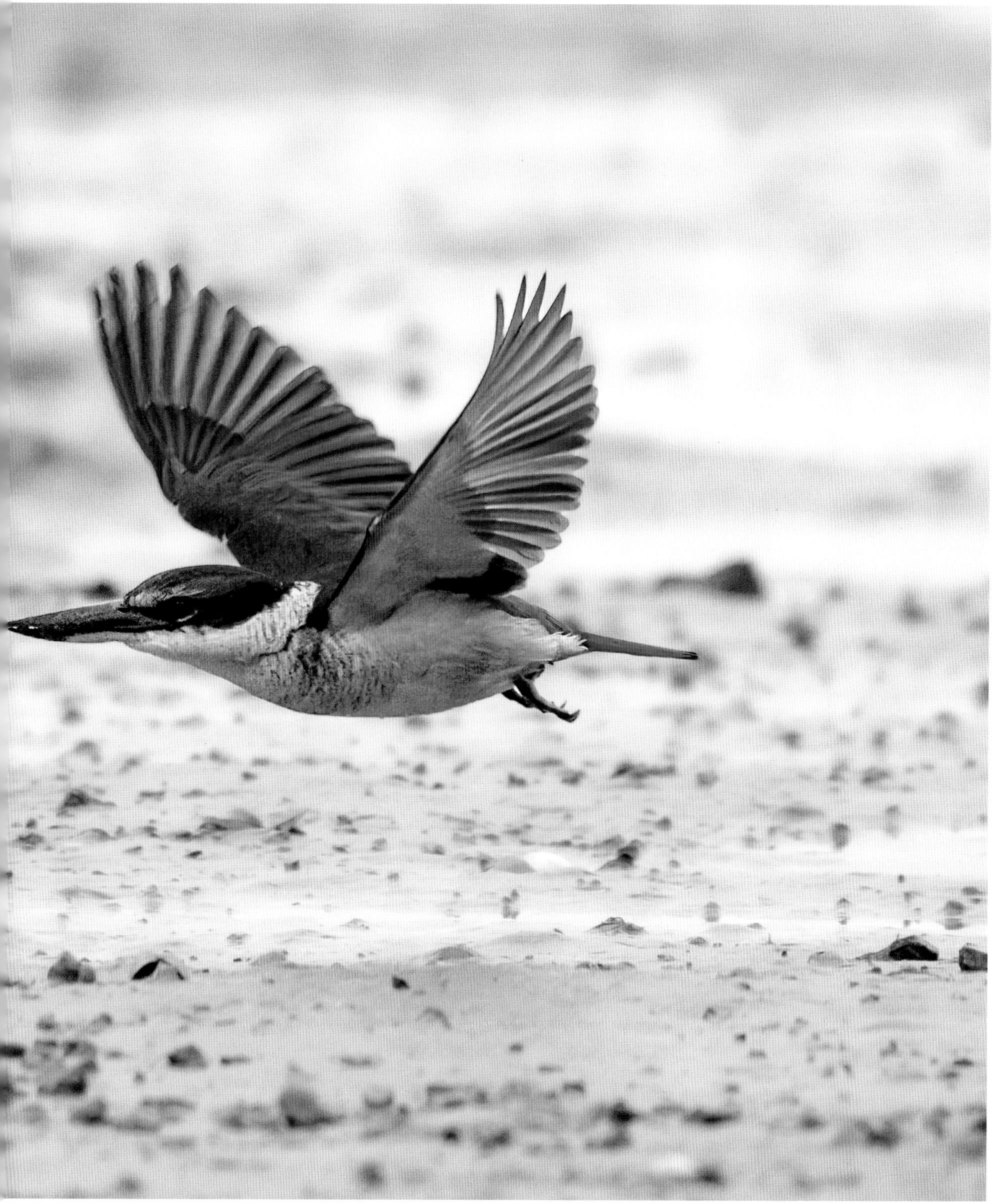

Kiwi

Mammal-like bird with more oddities than just its knitting-needle beak

North Island brown kiwi

Māori name	Latin name	New Zealand status	Conservation status
Kiwi	*Apteryx mantelli*	Endemic	Declining

Little spotted kiwi

Māori name	Latin name	New Zealand status	Conservation status
Kiwi pukupuku	*Apteryx owenii*	Endemic	Recovering

Great spotted kiwi

Māori names	Latin name	New Zealand status	Conservation status
Roa, roroa	*Apteryx haastii*	Endemic	Nationally vulnerable

Ōkārito brown kiwi

Māori name	Latin name	New Zealand status	Conservation status
Rowi	*Apteryx rowi*	Endemic	Nationally vulnerable

Southern brown kiwi

Māori name	Latin name	New Zealand status	Conservation status
Tokoeka	*Apteryx australis*	Endemic	Nationally endangered

The kiwi is said to be an honorary mammal: this strange, shaggy, bumbling figure evolved to fill the role taken in other parts of the world by a hedgehog or anteater. With its tailless rump, and its useless little wings tipped with 'cat claws', it spends its nights scurrying here and there over the forest floor, crashing through the undergrowth, snuffling and grunting, and probing the ground with its long, straight beak.

It has feathers like hair, and instead of regular hollow bird bones, it has heavy, marrow-filled bones like those of a mammal.

The kiwi's extraordinary beak is its greatest asset: with nostrils right at the tip (rather than at the base, like other birds), the bird is equipped with an incredible sense of smell. The beak has been compared to a 'blind man's walking stick': the kiwi forages by touching it to the ground for a few seconds here and there, before running off and touching the ground somewhere else, snuffling and snorting as it goes. In 2007, the research group of kiwi scientist Isabel Castro discovered a special organ in the last 5 mm of the beak. It is made up of pitted bone, and in each little pit there are receptor cells. Touching the beak to the ground enables a kiwi to pick up the vibrations or pressure signals of worms under the ground. It then plunges the beak in and bags its prey. The pig-like snorting cleans the nostrils of dirt and also spreads nasal drip down the bill so that the kiwi can dissolve smells in it. This organ isn't exclusive to the kiwi; it evolved independently in some shorebirds, too, including the godwit.

The kiwi also has a great sense of hearing: its ear holes are large, and obvious, too; not hidden like other birds'. With hearing and smell so specialised, it doesn't seem to matter that the eyes are so tiny. Whereas most nocturnal birds have eyes adapted for low light (often the eyes are large and sensitive, as in owls), the kiwi has the smallest eyes compared with body mass of any bird. In fact, in 2017, a study on 160 individuals in the wild found a third of them had eye lesions, yet were healthy and thriving, demonstrating that kiwi are not dependent on their eyesight at all.

The kiwi has one of the largest preen glands of any bird. This gland — which looks a bit like a nipple — secretes an oil that the bird spreads throughout its feathers with its beak, keeping the plumage in tip-top waterproof condition; the scent may also be used in social recognition. The kiwi's preen gland has eight oil-producing parts below the skin — most birds with more than two are water birds, so it's thought the kiwi has grown this enormous gland to keep its feathers functional and warm in New Zealand's damp rainforest conditions. Also, because the feathers are hair-like (they lack the barbules that make them stiff like other birds'), it's possible they might need more oiling than usual to stay robust. Some kiwi love the water — the brown kiwi often roosts in 'water burrows' with its feet and belly on the water. Kiwi do not dust-bathe like other birds.

Kiwi can run as fast as humans — when early Māori were running after their enemies, sometimes they would chant words to evoke the speed abilities of the kiwi and the weka. Sometimes Māori hunted kiwi, but sparingly: only chiefs were allowed to eat the meat and wear the kiwi feather cloak, or kahu kiwi. Hunters would cut tracks in the bush and then lure the birds by imitating a male call, attracting either the female or the male,

The Stewart Island/Rakiura southern brown kiwi is the largest of all kiwi, and may be seen foraging during the day.

A young kiwi in the burrow with its father.
The remains of the massive eggshell can be seen just underneath its feathers.

who would be puffed up ready for a fight. They'd also often use dogs and traps to catch them. Kahu kiwi, made from flax fibre and kiwi feathers, are greatly treasured and are worn ceremonially as symbols of high birth and leadership. Kiwi feathers are still used in new cloaks to this day, obtained (with a permit from the Department of Conservation) from kiwi that died naturally or were killed by cars or predators. Early Pākehā explorer Charlie Douglas wrote about eating kiwi in the late nineteenth century, but was unimpressed: 'They have an earthy flavour . . . [like] a piece of pork boiled in an old coffin.'

There are five species of kiwi left. The North Island brown kiwi is found throughout the northern and central North Island. The little spotted kiwi is a small grey bird now extinct on the mainland but found on some offshore islands and at the Zealandia sanctuary. The great spotted kiwi lives in high-altitude tussock and forest in the north-western South Island. The Ōkārito brown kiwi is critically endangered in Ōkārito forest. The southern brown kiwi or tokoeka has populations in Fiordland, Haast and Stewart Island/Rakiura: its Stewart Island subspecies is the largest of all kiwi, and may be seen foraging during the day, sometimes visiting the beach for sandhoppers.

The species all look different and also vary greatly in behaviour. In North Island brown kiwi and kiwi pukupuku, the male

incubates the eggs all by himself, but in the other species the female and helper-adults might also assist. Also, the North Island brown kiwi can lay up to six eggs in a season (and kiwi pukupuku, four), whereas the others usually lay only one.

The kiwi excavates a burrow (or chooses an existing one) and lays a massive egg that is up to 25 per cent of the adult's body weight. The kiwi has two functional ovaries — almost all other birds have one functional and one useless (it's suggested they did this to shed weight for flight). If the kiwi lays more than one egg, its ovaries take turns at ovulating.

The egg is incubated between 65 and 90 days; this is twice as long as most birds. Some early Māori believed kiwi took two or three years, and it was thought they abandoned their eggs until the chicks hatched — because in species where the male incubates alone, he must leave the egg unattended for much of the night while he forages, kicking leaf litter over the egg to conceal it. When it does hatch, each chick kicks its way out of the egg (it has no egg tooth), which is exhausting and can take a couple of days. In these first few days of life, neither parent feeds the chick, which is nourished instead on a bellyful of yolk, having eaten the remainder of the yolk before hatching. (The jumbo kiwi egg is almost 65 per cent yolk, versus 35-40 per cent in most birds' eggs). Within hours to a day it is ready to start foraging for itself on the forest floor at night.

Kiwi are the smallest living ratites, and the only nocturnal ones. Other ratites are the ostrich, emu, cassowary and rhea; all are birds that have become flightless and developed powerful legs. You'd be forgiven for thinking kiwi share their ancestry with New Zealand's extinct giant ratites, the moa. But their closest relative is, in fact, the extinct elephant bird of Madagascar — a giant, ostrich-like bird that stood 3 m high and weighed up to 730 kg, and died out three centuries or more ago. What's more, the kiwi ancestor was thought to have been a flying bird, not one that was flightless and drifted with the shifting continents.

Kiwi split into its different species during the Pleistocene epoch. In 2016 it was discovered today's five species descend from 11 lineages, and there were a further five or six lineages that are now extinct. It's thought that when the ice spread throughout New Zealand, kiwi populations were cut off from each other and started

Kiwi have poor eyesight and rely primarily on smell when foraging.

to subtly change — for example, their calls and smells became so different they were no longer attracted to other groups. This isolation and evolution happened over and over again as the ice expanded and shrank seven times in almost 800,000 years.

Kiwi used to exist throughout New Zealand in their millions, but forest and scrub clearance, along with introduced predators (cats, stoats, dogs, ferrets, possums, pigs) have laid waste to kiwi numbers. Without predator control, 90 per cent of chicks die before six months of age, only 5 per cent reach adulthood, and dogs and ferrets can even kill the adults. Kiwi numbers in areas managed by DOC and community groups are stable or increasing, but everywhere else they are in slow decline — there may be fewer than 70,000 left. Unattended dogs are a huge problem: they can easily crush kiwi to death with their jaws. In 1987, a dog that went rampant for two months in Waitangi Forest in Northland was estimated to have killed around 500 kiwi, which was half the population in the area. Many hunting dogs are put through kiwi aversion training to try to avoid altercations in the bush. The work of communities such as those behind the Backyard Kiwi project at Whangārei Heads has been instrumental in increasing kiwi numbers. Predator control, pest and weed control, kiwi tracking, kiwi releases, and campaigns on the dangers of dogs and cars have increased kiwi numbers from 80 in 2001 to now more than 500; they have also created a sense of connection to this bird in the community.

Lesser albatrosses

Meandering mollymawks

Light-mantled sooty albatross

Māori names	Latin name	New Zealand status	Conservation status
Toroa pango, toroa haunui, kōputu	*Phoebetria palpebrata*	Uncommon native	Declining

White-capped albatross/mollymawk

Māori name	Latin name	New Zealand status	Conservation status
Toroa	*Thalassarche cauta*	Native	Declining

Salvin's albatross/mollymawk

Māori name	Latin name	New Zealand status	Conservation status
Toroa	*Thalassarche salvini*	Endemic	Nationally critical

Chatham Island albatross/mollymawk

Māori name	Latin name	New Zealand status	Conservation status
Toroa	*Thalassarche eremita*	Endemic	Naturally uncommon

Buller's albatross/mollymawk

Māori name	Latin name	New Zealand status	Conservation status
Toroa	*Thalassarche bulleri*	Endemic	Naturally uncommon

Light-mantled sooty albatross

This is a Siamese cat in bird form, with a sooty brown or black head, dark wings and tail: arguably one of the most beautiful albatrosses. It was called the 'blue bird' by early sealers, because its plumage can look blue in strong Antarctic light.

The largest colonies are found on South Georgia, and the Kerguelen and Auckland islands: it also nests on other subantarctic islands and New Zealand has about 30 per cent of the species globally.

It nests every second year on steep slopes and cliffs, launching straight off into the air when the wind allows. The birds effortlessly glide and swoop on pointed wings, sometimes appearing to stand still in the air, preening their bellies even in flight, without flapping. They feed as far south as the Antarctic ice. Like the related sooty albatross, the light-mantled sooty albatross has a comical white crescent of feathers around the eye, but it has a unique light blue streak along its dark lower bill. Together these two features make the bird look eternally, happily surprised.

White-capped, Salvin's, and Chatham Island albatrosses

These birds are your classic average-sized mollies: black wings and back, and a long bill with a fierce hook tip. They were widely considered to be the same species (with subtle variations in bill colour and the patterns of grey on the head), but many experts have now split them up, because they breed in different ranges. Salvin's albatross breeds in large, busy colonies on the Bounty Islands and Snares Western Chain; the white-capped albatross nests on some of the Auckland and Antipodes islands.

The Chatham Island albatross has a beautiful grey head; it is quite rare (with fewer than 5000 breeding pairs) and breeds only on The Pyramid, an inhospitable rock stack in the Chathams.

Chatham Island albatross chicks were once commonly taken by Moriori for food, feathers, and bones for tools. Moriori rafted over to the stack on waka kōrari or wash-through rafts (there were no large trees

LEFT
Storm-riders of the Southern Ocean, Salvin's albatrosses — like all albatrosses — are perfectly suited to life in the Roaring Forties latitude.

RIGHT
Chatham Island albatrosses breed only on The Pyramid in the Chatham Islands archipelago, although chicks have been translocated to a sanctuary on the main island in the hopes of securing another safe breeding colony.

there to make dug-out canoes), but in later years, when people started using European boats and taking thousands of chicks, this practice became unsustainable.

All three albatrosses are caught and killed as bycatch in long-line fisheries, both in New Zealand waters and in their far-flung feeding grounds off Chile, Peru, South Africa and Namibia. And of course, as with all seabirds, mistaking floating plastic trash for food is an increasingly major killer, with rubbish now found in albatross colonies even on remote, otherwise pristine, subantarctic islands.

Buller's albatross

Buller's albatross is one of the smallest albatrosses and breeds only on New Zealand offshore islands. Very strangely for an albatross associated with rocky open colonies, the southern subspecies often breeds deep in forest, under the dense, woody tree daisies and hebes of the Snares and Solander islands. Sometimes these birds have to waddle through the forest 100 m inland to reach their nests. The northern subspecies nests on more typical albatross terrain: the cliffs and patchily

herb-covered hills on the Chatham and Three Kings islands. Just as with the northern royal albatrosses, a fierce storm in the 1980s on the Forty-Fours in the Chathams group destroyed the northern Buller's nest-building vegetation and knocked the population back.

During the breeding season, Buller's albatrosses are often seen scavenging close to fishing boats around the coasts south of Cook Strait (sometimes to their detriment: deep-sea longliners and trawlers are the main threats to this species). The birds tend not to plunge or dive, instead feeding from the sea surface on fish, squid, krill, salps and offal. Outside the breeding season, they're off, wheeling and soaring effortlessly to the seas off Peru and Chile. Once the single chick fledges, it doesn't come back to breed for about 12 years. Another name for this bird is Buller's mollymawk. Used only in New Zealand, 'mollymawk' derives from the Dutch words for 'foolish gull', and is usually reserved for any albatrosses in the genus *Thalassarche*.

There are two subspecies of Buller's albatross: a northern, which breeds on the Three Kings and Chatham islands, and a southern, which breeds on the Snares Islands.

Morepork

Wide-eyed hunter of the night

Māori names	Latin name	New Zealand status	Conservation status
Ruru, koukou, peho	*Ninox novaeseelandiae*	**Native**	**Not threatened**

The ruru deftly manoeuvres through dense New Zealand bush at night, its short wings incredibly responsive and also silent: like other owls around the world, it has serrated edges on its flight feathers, which muffle the sound of air passing over the wing.

Along with its partner, it'll have a territory of about 3-5 hectares over which it has an intimate knowledge of what small bird is nesting where, and it will prowl around waiting for a chance to grab some tasty chicks for dinner. An owl's eyes cannot move in their sockets, and so instead, the ruru can quickly swivel its head up to an incredible 270 degrees without moving its shoulders. The eyes are adapted for the dark, with most of their photoreceptors specialised for low-light vision rather than for colour vision. Hearing is super-sensitive, too, with the facial discs (two depressions made by skull shape and feathers) funnelling all sounds towards the ear openings. The ruru zeroes in and grabs prey with its talons, either striking it out of the air or seizing it on the ground, and then holds its victim with one foot while tearing off wings or mouthfuls of flesh with its hooked beak.

Although the morepork does take small birds, mice, rats and lizards, most of its food is insects (such as cicadas, wētā, moths and beetles) and spiders. If there are any indigestible bits, such as bone or exoskeleton, it heaves them up later in pellets, which pile up under a much-used perch.

This speckled brown owl, with its dark mask and staring yellow eyes, is New

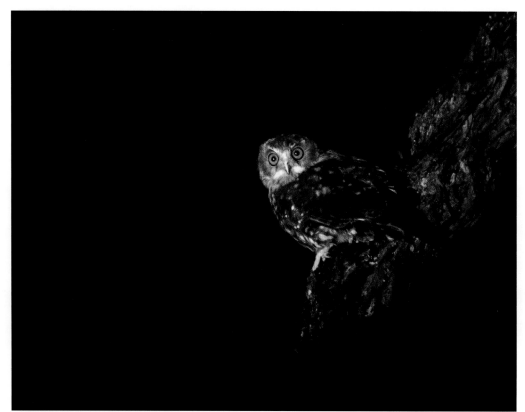

As well as having their typical *ru-ru* call, ruru can also sound like crickets buzzing, or even kiwi calling.

Zealand's only surviving native owl; a larger bird, the whēkau or laughing owl (which sounded like a laughing madman), became extinct in the early nineteenth century. Australia has a subspecies of the morepork, called boobook or mopoke. Ruru are thought to be declining, but are still relatively common in most of the country (apart from eastern and central South Island) — living anywhere from your suburban park to the deepest native or exotic forest, and from the coast to the upper bush line. They are sometimes seen outside windows at night, spookily staring into the house while they watch for moths fluttering around the light, catching them always in their talons. Their *more-pork* duet is one of New Zealand's iconic night sounds, although rather than duetting lovers these could also be the calls of rival males having a verbal duel. Ruru also have a *quork-quork* call and a *quee*, which rises and is often mistaken for a kiwi. When they're displaying to defend their territory or nest, they puff up their feathers and hold their wings slightly out to appear impressively bulky.

Fearsome predators by night, ruru are sleepy puffs of feathers by day — dozing on their daytime perches in deep shade. If stumbled across by smaller birds, such as bellbirds, fantails or silvereyes, they can be mobbed noisily, harassed until they head elsewhere to finish their snooze.

They make their nest anywhere cosy and dark — in a tree hollow, among epiphytes on a branch, on the ground under tree ferns or even in a seabird's burrow. When courting, they bow, wing-flap and sway, and the male often offers the female food. He'll also do this before she lays, while she's developing the eggs inside her, while she incubates and in the early days of the fluffy grey chicks' lives. The parents hunt to find food for the chicks, but eventually they'll cut the apron strings and aggressively push the young out of their territory.

Among many early Māori tribes the ruru was a bad omen if its sound was heard during the day, and even more so if it was ever found inside a house. Being a night bird, it was believed to have come from the underworld, the realm of the dead, and some believed it had a taste for eating dead men's fingernails (which it incorporated into its eyes, which were never eaten by Māori — indeed, some tribes never ate ruru at all). This is similar to the Australian Aboriginal belief that a sighting of their mopoke is a forewarning of death.

Other Māori tribes, however, saw the ruru as a kaitiaki or guardian, and they would carefully listen to its calls. Author Murdoch Riley wrote that its glaring, unblinking eyes may have been the inspiration for the wide eyes in the pūkana facial expression of the haka.

Among those tribes that did eat ruru, one method of hunting it was to 'hypnotise' it — the hunter would perform the actions of the haka, fascinating the bird, while edging close with a slip-noose.

While ruru eat rats and mice, the introduction of predators to New Zealand has been bad for the owls. Cats, possums and rodents eat their chicks and eggs in the nest (and pigs and hedgehogs take them on the ground), and they are competition for food. When poison is used to kill rats, some morepork may get secondary poisoning from hunting and eating the poisoned rat; however, research has shown morepork numbers in poisoned areas overall will seesaw up as numbers of their predators and competitors go down.

Caught by
day, ruru are
squinty and
snoozy.

New Zealand dabchick

Sleek lake-dweller that ferries its chicks

Māori names	Latin name	New Zealand status	Conservation status
Weweia, totokipio, taihoropi, taratimoho	*Poliocephalus rufopectus*	Endemic	Recovering

With fine silver feathers swept back suavely over its head and pale yellow eyes, the New Zealand dabchick can be seen bobbing about on lakes and farm ponds throughout the North Island and in parts of the South — with the occasional call of *wee-ee-ee* (hence the Māori name weweia). It's New Zealand's only endemic grebe, and with habitat loss and the toll from introduced predators the bird has declined over the years.

When alerted, the dabchick extends its neck like a slender periscope, bum fluffed up. Then, if the danger's real, it might suddenly dive under the surface, sending a jet of water up into the air with its powerful feet. Like other grebes, the dabchick can control its own buoyancy — trapping air in its dense, waterproof feathers to float, then compressing these feathers against its body to sink further into the water, leaving only its head and neck visible.

The dabchick breeds from August to March, kicking off with feisty territorial battles (involving lots of rushing at opponents) and outrageous courtship displays — these involve side-to-side head-flicks, diving under each other, pattering and splashing their feet, preening, shaking

LEFT
Dabchicks will display and chase interlopers, like this hoary-headed grebe (left), out of their territory, Lake Elterwater.

RIGHT
As adults, dabchicks have startling white eyes.

and rearing up out of the water.

The nests are messy, waterlogged mounds of grass, reeds and sticks. The birds anchor these 'compost heaps' to trailing tree branches or marginal plants; other favoured spots are caves and crevices. Unfortunately they don't seem to choose materials that float, so the nests cannot ride any rising floodwaters or boat wakes. Wakes are, in fact, one of the major threats to dabchick eggs and chicks.

During the first three weeks of life, chicks sit on the back of an adult, and like the pedal-swans in tourist lakes they will ride their parents across the water, the adult's wings slightly lifted to keep them from falling off. The riders themselves look like scruffy avian tiger cubs — the chicks' fluffy head plumage features black-and-white horizontal stripes and sometimes an orange crown.

Dabchicks eat insects, such as midges, waterboatmen and dragonflies, snapping them from the air or picking them off the water's surface. They can also dive to 4 m for up to 40 seconds to grab small native fish like bullies, as well as freshwater crayfish, shellfish and leeches.

New Zealand falcon

High-speed fighter-bomber

Māori names	Latin name	New Zealand status	Conservation status
Kārearea, kārewarewa, kaiawa	*Falco novaeseelandiae*	Endemic	Recovering

The New Zealand falcon or kārearea is a high-speed aerial acrobat, tearing through the air at speeds of more than 100 km/h, swooping and careering around trees after live prey. It can see eight times better than humans, with incredibly large eyes for a bird its size. It's the most aggressive falcon in the world when defending its nest, dive-bombing nearby humans and animals alike with fisted feet, and raking talons ferociously across heads.

Of New Zealand's four native birds of prey (the falcon, swamp harrier, morepork and, recently, the barn owl), the kārearea alone is endemic: found here and nowhere else in the world.

How can you tell it apart from the kāhu or swamp harrier? For one, it's a lot rarer than the commonly seen kāhu. Physically, the kārearea is two or three times smaller — only about the size of a magpie. Its eye is large and dark, rather than yellow, and it doesn't have the fierce eyebrow ridges of the kāhu. The kārearea doesn't indulge in roadkill, either, preferring to chase and attack fresh live prey. And while other falcons have long, narrow crescent-shaped wings for fast flight in open country, the kārearea's wings are short and rounded;

Unlike the languid soaring of the kāhu, the flight of the falcon is rapid and direct.

this, together with its long tail, perfectly suits it to manoeuvring tightly through dense forest at great speeds on the hunt.

And a-hunting the kārearea goes, sometimes by contour hugging — flying silently above the bush canopies or close to the ground at speed, taking fantails and skylarks by surprise as it happens upon them. Other times it'll perch on trees or power poles, seeming to innocently preen itself but suddenly exploding into action and launching an attack. Or it'll soar up to 1000 m in the air, intercepting other birds in flight or tucking its wings in like a dart and diving insanely fast to smash its prey. If the prey escapes and hides under bushes, tough luck — because the falcon will walk around until it flushes it out.

The kārearea seizes prey with long, sharp talons on long toes, perfectly designed to plunge deep into feathers or fur. Like other falcons, its fierce beak has a tomial tooth, a thorny-looking tab on the cutting edge of the upper beak, with which it swiftly severs a victim's spinal cord. A quick kill is important, given that prey (such as hares, poultry or the white-faced heron) may be six times the falcon's size, and any struggles could seriously injure the falcon. If the prey is a small bird, the kārearea will fly with it to a perch to pluck and devour it.

Though the kārearea is just one species, South Island birds are larger than those in the North Island. Always, the female is much bigger than the male. She lays her dark mottled reddish-brown eggs on the ground in all types of places: tussocklands, roughly grazed hill country,

Found throughout New Zealand from alpine to lowland and even subantarctic habitats, the kārearea is still our rarest raptor.

pine plantations and dense native forest. They nest in small scrapes under logs or bushes, on sheltered ledges on cliffs, in tree hollows or among epiphytes (such as astelias) growing on tall trees, such as rimu or kahikatea, that poke up above the forest canopy.

Unexpectedly, large exotic pine plantations can be great homes for falcons, so long as only parts are felled at a time. The falcons nest in scrapes in the felled areas where there's a lot of unmilled logs and branches. The fledglings can safely learn to fly by climbing up the logs and launching off, and the parents can still use nearby forest stands as vantage points and hunting grounds.

The kārearea's main call is a high-pitched, rapid, almost ear-piercing *kek kek kek*, heard when defending a territory — with the female's voice deeper than the male's. Adults also chitter amongst themselves, especially after mating. Sometimes the young and the adult females will whine for food.

In Māori mythology, the kārearea stars in a love story and the naming of Whangārei — reflected today in murals

High vantage points are
good for spotting small
birds to dive down onto.

by artists Charles and Janine Williams in
Hamilton and Whangārei. Twin sisters
Reitū and Reipae lived in the Waikato,
and a handsome young chief up north
in Whāngāpē wished to marry them. He
sent a messenger in the form of a kārearea
(or at least, someone that moved like a
kārearea) to pick up the twins. But after
an altercation, Reipae stayed en route
in Whangārei to wait for her brothers to
arrive and eventually married another
chief, while the other twin headed up
north. This is the story behind one of
Whangārei's longer names: Whanga-a-
Reipae (the harbour of Reipae).

The kārearea is classed as at risk; the
Department of Conservation estimates
numbers as between 5000 and 8000, but
this number is uncertain. Threats include
loss of nesting and hunting habitat,
introduced predators (such as cats and
mustelids, which raid the ground nests),
and humans who shoot, trap or poison
falcons to protect their domestic birds.
Electrocution by power lines is another
big killer — implicated in almost half of
all falcon deaths in one Marlborough
study of 55 birds — especially in open
areas where the lines are among the only
perches around, substituting for trees
from which to swoop onto prey.

One of the falcon's staunchest allies
is the Wingspan National Bird of Prey
Centre, based in Rotorua, which works
to conserve New Zealand's raptors. The
outfit is operated by Debbie Stewart and
Noel Hyde, plus a small army of staff and
volunteers. They rescue, rehabilitate and
release the birds, and they also run an
outreach programme to educate visitors
and the wider public about these often
maligned hunters. Helping the birds can,
for instance, involve mending feathers.
The kārearea's soft feathers are perfectly
adapted for forest life — because they're
flexible, they won't snap off as easily in the
bush environment. But they still can snap,
so the crew at Wingspan will create feather
extensions by gluing spare feathers into
the base shafts of broken flight feathers.
This technique, from an ancient falconry
practice called imping, enables birds to
keep flying while their new plumage grows.

New Zealand fantail

Fidgety, friendly forest follower

Māori names	Latin name	New Zealand status	Conservation status
Pīwakawaka, tīwakawaka, tīwaiwaka, tīrairaka, tītakataka	*Rhipidura fuliginosa*	Endemic	Not threatened

A kissing sound in the forest is one of the first signs you've got a fantail nearby — a squeaky kind of *cheet cheet* chatter from high on a perch — and soon, this tiny fluttering bird will be following just a metre or so behind you, now giving a *pip* hunting call. But it's not you it's after: the fantail's aerobatics are just a way of snapping up the insects you disturb as you go.

The black, white and buff fantail, named flycatcher by early Europeans, has at least 20 names in Māori, including tīwaiwaka, pīwakawaka, tīwakawaka, and tīrairaka, many speaking to its energetic, never-resting nature. One of its names, tītakataka, is used to describe someone who just can't stay still: he tou tītakataka.

The pīwakawaka's tail makes up more than half the length of the bird, and allows it to pivot its tiny body in flight, twisting and turning super-quickly to catch prey on the wing. It is a connoisseur of small insects (moths, flies, beetles) and spiders, gobbling them up three ways. First, there's hawking, where it spies a bug from a perch then snaps it up in flight; next, flushing, in which it ruffles up dense, leafy trees or bushes and grabs the bugs as they fly out; and finally feeding in association, where

Paradoxically, fantails may be more common now than they were before humans arrived with their mammalian predators.

it simply follows other birds (or any large animals, including humans) to snap up the bugs disturbed. If they take large prey, they hold it in one foot against a perch and peck it again and again to kill it.

Pīwakawaka are found everywhere from Northland to Stewart Island/ Rakiura, as well as some of the Chatham Islands — anywhere there are lots of trees, from native forest to parks to urban gardens. They don't do too well in the cold, however: long cold and wet spells can wipe out local populations (which soon recover because these birds breed so fast). Paradoxically, fantails may be more common now than they were before humans arrived with their mammalian predators, thanks to the reduced competition for food. A study by

Colin Miskelly looked at forest sanctuary Zealandia in Wellington, a predator-free area that has had deeply endemic (i.e. birds that have been endemic a very long time) forest birds reintroduced. Since the endemics have become abundant, three recently endemic or native birds (which are less specialised for New Zealand's ecosystem) have declined hugely: these are the grey warbler, silvereye and fantail. It could be that the deeper endemics — which include the tūī, stitchbird and North Island robin, among others — are very good at gleaning insects off twigs, leaves and bark in this particular ecosystem, and so the fantail no longer has such easy and abundant food.

Pīwakawaka vary in colour. The most common morph is the pied fantail; its

Fantails are common in many habitats, from forests to wetlands.

black-and-white tail can be seen fanned out and flushing prey all over New Zealand. In the South Island, five to 25 per cent of the birds are black or dark brown with a dark fan; this so-called black morph is very rare in the North Island. It's not known what the advantages of each morph are; it may be that the black tails are harder wearing than the pied fantail's feathers (with birds, more melanin in feathers often means less wear and tear), yet the pied colour is possibly better for flushing out prey. Regardless, the two morphs will happily couple up. While they're usually seen in pairs or alone during the breeding season, in a very cold winter (for example, in the deep south) they will sometimes huddle together in roosts of more than 20 birds to conserve heat.

The female weaves a compact, cup-shaped nest from all sorts of materials, such as grasses, wood, hair, moss and spider webs. Fantails are very territorial during the breeding season, and for good reason: these chick-making machines can build up to four or five nests in one season, with up to five eggs in each. Chicks stay in the nest only two weeks, and by then the female is already building her next nest. Introduced predators raid fantail nests, as do moreporks, which especially like the plump, week-old nestlings. Ship rats are the bird's worst enemy. The most common type of rat in New Zealand, the

LEFT
Bristles, framing the beak, funnel hapless insects into the mouth of the fantail.

RIGHT
Black morph fantails are more common in the southern parts of New Zealand.

ship rat habitually climbs trees and will eat eggs and chicks, plus any adult that tries to protect its nest. Thankfully, fantails manage to maintain their numbers by being such great breeders.

In Māori lore a pīwakawaka that enters a home is seen as a forerunner of death — something not necessarily to be feared, but an opportunity to prepare for the event. This is because the fantail was behind the death of the demigod Māui. He wanted to gain immortality for humankind by destroying Hine-nui-te-pō, the goddess of the night and underworld, who had eyes of greenstone, hair of sea-kelp and a mouth like a barracouta. In one version of the myth, Māui took some small forest birds

with him — including the tomtit, grey warbler, robin and fantail. He found the goddess sleeping, her legs wide apart and sharp spikes of obsidian and greenstone between her thighs. He planned to pass through the goddess's body, in through her vagina and out of her mouth. This would kill the goddess and let humans live forever. He managed to get his head and arms in, but this looked so ridiculous the pīwakawaka couldn't hold back its laughter (in another version, Māui transformed into a worm to get in). In all versions, the goddess woke and crashed her thighs together, the flints cutting Māui in two and killing him.

New Zealand pipit

The one with the waggly tail

Māori names	Latin name	New Zealand status	Conservation status
Pīhoihoi, hīoi, whioi	***Anthus novaeseelandiae***	**Endemic**	**Declining**

That tail. It bobs up-down, up-down, as the endemic New Zealand pipit hurries along the ground with the odd pause to look around. The bird looks like a streaky grey-brown sparrow with a slightly longer tail — and a nervous tic.

Why all this wagging? It is in fact a characteristic of the whole family of wagtail birds around the globe. It may serve to flush out insects — or to show potential predators that the pipit is vigilant and alert, and therefore not worth pursuing. Black phoebes in California, birds that also wag their tails, do it three times more when they hear the cry of a hawk than when a harmless bird calls. The pūkeko similarly will flick its tail more frequently when on alert. Still, you can get quite close to a pipit. Although it can fly, it will just run along the ground to move away.

The *tzweep* of the New Zealand pipit may be heard anywhere in wild open country — from the seashore, riverbeds, rough pasture and road verges up to the subalpine tussock at around 1900 m.

Walter Buller wrote in 1888 that it can go higher — even to 'the very summit of Mt Egmont'. Pipits also move into freshly felled areas of plantation forestry. When Māori first arrived and burnt off swathes of native forest, the pipits probably benefited. Pākehā settlement would have affected numbers, too — certainly from the 1880s till the 1920s they were present in huge numbers. Herbert Guthrie-Smith of Tūtira described flocks in their hundreds. But they're little seen on developed farmland, such as that of Auckland, Waikato, Bay of Plenty and Canterbury, or on the developed areas of the central plateau, probably owing to introduced predators and intensified land use.

Pīhoihoi in Māori, it was named for its call: hoihoi means 'noisy' or 'be quiet'. The pipit is also called whioi, 'whistler'.

Pipits are found in abundance on our remote subantarctic islands, as well as on the Chatham Islands.

The name pīhoihoi was used as the Māori translation of sparrow, when in 1863 the New Zealand government set up a Māori newspaper to rival one already existing in the Waikato. They named the press *Te Pihoihoi Mokemoke i Runga i te Tuanui*, from Psalm 102, verse 7 in the Bible ('I watch, and am as a sparrow alone upon the house-top').

Pipits are almost never seen in trees, preferring to perch on stumps, logs, fence posts and rocky outcrops. They gobble up seeds and fruits, but mostly they're after insects — anything from beetles to crickets to wasps. They're even partial to sandhoppers on the seashore. In fact, the bird first became known to Europeans in 1773, when naturalist Georg Forster and his father, who were on James Cook's second voyage, saw the pipits feeding on small crustaceans in the seaweed on the shore of Queen Charlotte Sound, and later shot some on Long Island. They called them 'new larks' and 'sand larks'. The pipit family occurs around the world, but the four New Zealand subspecies occur nowhere else — the New Zealand, Chatham Island, Auckland Island and

Pipits are said to be the second most abundant avian food source for the falcon.

Antipodes Island pipits. Unlike their overseas counterparts, our endemic pipits don't seem to migrate or move long distances.

Pairs will stay in the same place year after year, which is where they'll also breed, raising two or three clutches per season. They make their nest — a cup of woven grass — on the ground partly hidden in a tussock or fern, or on a ledge on a cliff. The parents always take a sneaky, concealed route back to the nest: they'll land 5–10 m away and walk to it. The young, which are fed for weeks until they are able to fly, have extremely loud begging calls that unfortunately attract stoats, rats and falcons. The fledglings, too, make easy prey for falcons, as they're very weak fliers until their tails grow long enough for flight control. In fact, pipits are said to be the second most abundant avian food source for the falcon, and were also a favourite snack of the now-extinct laughing owl.

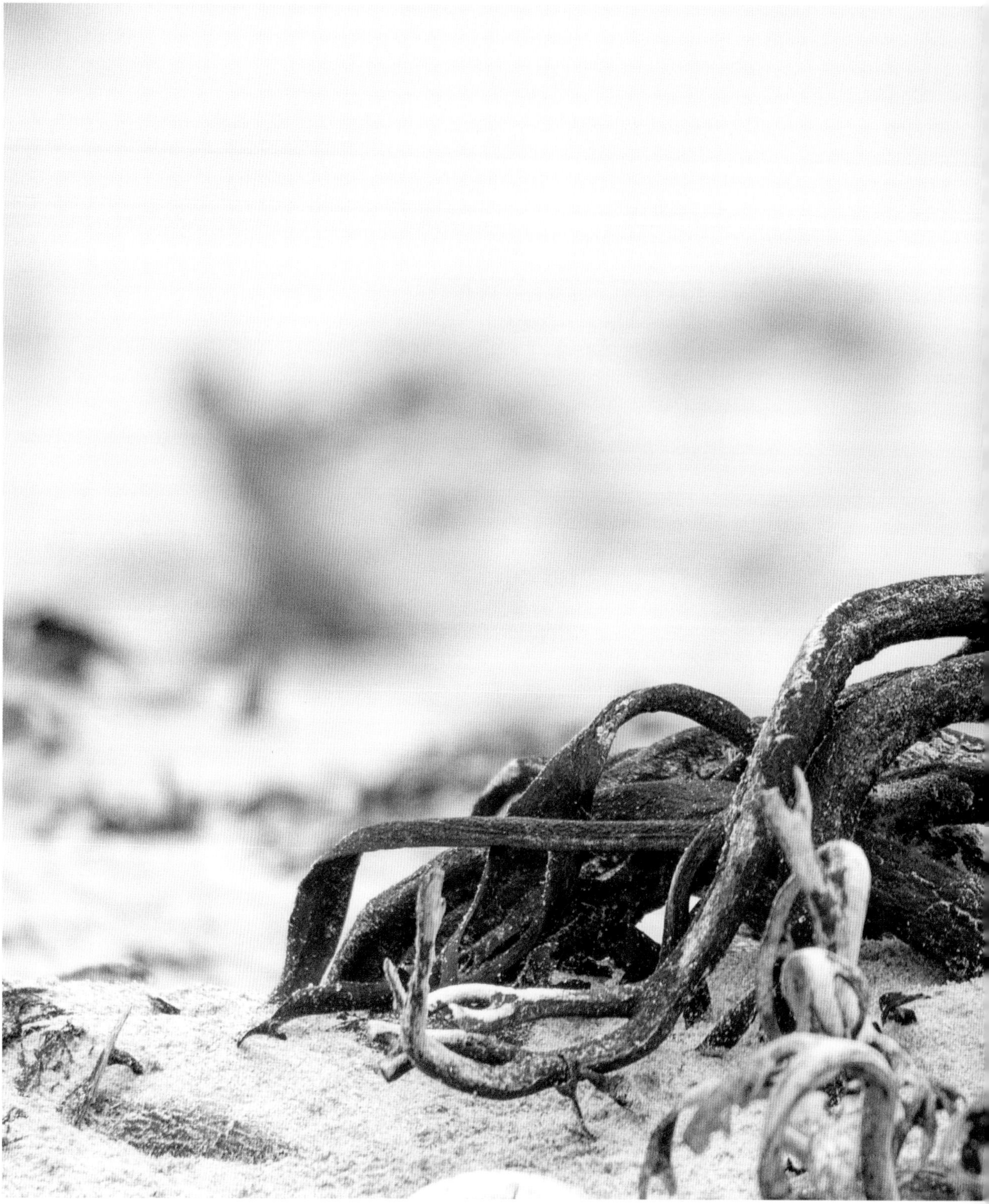

Pipits are often found foraging among washed-up kelp on the beach, Chatham Island.

North Island kōkako

Heavenly songster

Māori name	Latin name	New Zealand status	Conservation status
Kōkako	*Callaeas wilsoni*	Endemic	Recovering

With blue wattles and a lone ranger mask, the North Island kōkako is a bit like a smoke-coloured crow at a dress-up party — but then it opens its mouth and the most gorgeous un-crow-like sounds pour out.

The call is a hauntingly beautiful, slow flute-like chime that permeates the forest. Both male and female undertake a duet throughout the day, kicking it off with a spectacular dawn chorus. The eerie call is used often in media to evoke the spirit of the New Zealand forest; it's played in the arrival area of Auckland Airport (along with cicadas, sheep and crashing waves), and it is peppered throughout wilderness-inspired movies such as Jane Campion's *The Piano*.

The kōkako also makes it onto the New Zealand $50 banknote. The picture is accompanied by a sky-blue mushroom, *Entoloma hochstetteri*, that was called werewere kōkako by the Tūhoe people — literally 'the kōkako's wattle' — because of the similar colour between the two. According to Tūhoe legend, the kōkako gets its blue wattles from rubbing its cheeks against the mushroom.

There are now just over 1500 pairs of these precious birds, up from a low of 400 in 1999. The North Island kōkako once had a South Island counterpart. It had brilliant orange wattles rather than blue. It was declared extinct in 2007, but then in 2013 it was reclassified data deficient — some believe it still exists somewhere in the west of the South Island or on Stewart Island/Rakiura. Believers call it the 'grey ghost', and the last definite sighting was in 1967 in Fiordland. The South Island Kōkako Charitable Trust offers a $10,000 reward for evidence of its existence.

The song of the kōkako has been lost from many of our forests, but efforts are being made to translocate them back into habitats that are heavily managed for pests.

Sometimes, it's not a female he's wooing.

The songs of the kōkako — the beautiful whistle, and its clicks and coos — are versatile: they attract partners, they are used to communicate with a partner once they've set up a year-round territory (males and females often duet together), and they're used to communicate with neighbours across their boundary. Their song also keeps evolving throughout their lifetime. Each bird will be developing its own unique twist on it, and even one population can have a few dialects. The dialect on Tiritiri Matangi is much more basic than some others; local kōkako team leader Morag Fordham calls it 'grunty' rather than flutey, and says maybe this was because the island's first kōkako were translocated as juveniles and had no one to imitate.

Unfortunately, kōkako often ignore other kōkako whose dialects they can't recognise, and this can make it quite tricky to mix populations. Luckily, the occasional kōkako will take a liking to an exotic partner — and because they're evolving all the time, translocated kōkako often adapt their dialects anyway.

Aside from serenading her, when trying to woo a partner, the male will break out his 'archangel' pose — he'll fan out his wings and tail and bob on the spot. Then, sometimes twigs are passed between the two birds.

But sometimes, it's not a female he's wooing. In the late 1980s and 1990s, some populations of kōkako were just not breeding. They were coupling up and defending their territories, but very few

Tiritiri Matangi is an easy place to see the elusive kōkako.

eggs were produced. It turned out, because so many females were killed by predators, most of the pairs were male-male (the sexes are hard to tell apart) — one of these pairs even made a nest. However, as soon as effective predator control allowed the few remaining genuine breeding pairs to breed successfully, the new young females turned the heads of the male-male bonded couples and broke them up.

Even though their English name is the blue-wattled crow, kōkako are not related to crows proper. The kōkako's wings are too weak to fly upwards — instead they hop up the canopy, using their long, thin black legs as springs and their wings just for balance, squirrelling through the branches, and then launch themselves out into the air to glide down over long distances on their short wings. Pairs move through the forest together, with the male usually leading the way, nibbling on fruit and leaves, sometimes buds, and taking insects too. They will often hold food with their feet, like parrots. One researcher described kōkako eating a case moth by holding it in one foot and squeezing down the case with its bill, like toothpaste. Along with kererū, the kōkako is the only bird that can open its mouth wide enough to eat — and so disperse — the fruits of some native trees such as tawa, miro and mataī. It also acts as a pollinator by feeding on the nectar of some trees.

Through its feeding habits, the kōkako helped halt the logging of mature native trees in the Pureora Forest in the Waikato in 1978. Since the 1940s, the forest had

Kōkako use their feet when they eat, plucking flowers or fruits with their bill and then delicately holding them with a claw.

been the site of one of the biggest logging operations in New Zealand. Protestors against logging climbed up tōtara trees, bringing on a moratorium; research later discovered the kōkako were reliant on food from vines and perching plants in the giant tōtara, and so eventually the logging was ended for good.

North Island kōkako are in 22 scattered populations, just 11 of them natural and the rest having been translocated. Unlike other members of the Callaeidae family — the saddleback, the now-extinct huia and the South Island kōkako — they were never completely driven off New Zealand's mainland by introduced predators. It was a close-run thing, though: female kōkako have pecking and wing-flapping fights almost nightly with ship rats trying to enter their nests. If it's early in the incubation stage, the poor mother will often just give up and let the rat take the egg, or sometimes the flightless chicks will leap from the nest to escape: some die when they hit the ground, but the survivors will be fed down there by their parents. Possums and harriers also go for the eggs, chicks and adults.

When pests are controlled, kōkako hatch one to three chicks; their wattles are like tiny concave pink flower petals that gradually take on the blue hue as they grow (even in the absence of blue mushrooms!). After a year, they're booted from the territory to set up their own — to attract a mate with their lone ranger mask, their wattles and their beautiful song.

Oystercatchers

Pied (or not so pied) piper that can shuck like a pro

Variable oystercatcher

Māori names	Latin name	New Zealand status	Conservation status
Tōrea tai, tōrea pango	*Haematopus unicolor*	Endemic	Recovering

South Island pied oystercatcher

Māori name	Latin name	New Zealand status	Conservation status
Tōrea tuawhenua	*Haematopus finschi*	Endemic	Declining

Chatham Island oystercatcher

Māori name	Latin name	New Zealand status	Conservation status
Tōrea tai	*Haematopus chathamensis*	Endemic	Nationally critical

Oystercatchers have mastered the look of being 'ticked off'. If another oystercatcher even looks like it's going to enter a pair's territory, the pair has a ritual all lined up: the fearsome territorial piping display. The two get in position: shoulders meanly hunched up high, bills almost touching the ground, wings slightly away from the body — and march!

Uttering shrill, penetrating cries, they chase the interloper or pace up and down their boundary looking mean and contorted. These long piping cries, common to oystercatchers around the globe, also serve in other, less tense situations, such as showing off to other members of a flock.

There are at least eight species of oystercatcher worldwide. New Zealand has more than its fair share, with three endemic species: the variable oystercatcher (VOC), the South Island pied oystercatcher (SIPO) and the Chatham Islands oystercatcher (CIO).

Despite their name, oystercatchers eat more than oysters: mostly molluscs of all types, but also worms, crabs and sometimes small fish. They snap up food from the surface of estuaries and seashores, or probe down deep with their long beak. Opening a bivalve shell is child's play when you can shuck like an oystercatcher. This shorebird works

Chatham Island oystercatchers look very similar to South Island pied oystercatchers, but they're found only in the Chatham Islands.

its bill around deep in the sand or mud until it makes contact with a shell, then hauls it up to the surface. Positioning it on the sand, it inserts the flattened tip of its long orange beak through the hinge area, cutting the muscle that holds the shell together. *Bon appétit!* If a bivalve happens to be out of the mud and gaping, it's even easier — the oystercatcher just stabs between the shell halves and then levers its beak through 90 degrees to prise the halves apart. And if the shell is thin enough, the bird can punch a hole through it, choosing the weakest spot or where borers have already punctured the shell. Oystercatchers are also pretty good at smashing limpets sideways off rocks.

South Island pied oystercatchers take long breaks from the seafood diet, spending half of each year eating earthworms and beetle larvae in pasture. This is because they're the migrant oystercatcher: not for them are summers trying to raise chicks in sandy scrapes vulnerable to extra-high tides like the VOC. Instead, SIPOs spurn the coast and fly inland to braided riverbeds, farms, lakes and bogs to breed, mostly in the South Island. After fulfilling their parental duties, they then migrate back to beaches and estuaries around New Zealand.

Winter is when the socialising happens: oystercatchers form crowds (of more than 10,000 birds in some places) in the Manukau and Kaipara harbours, the Firth of Thames, the upper South Island and many other places around both main islands. Because SIPOs migrate inland but the VOCs stay put, winter is the only time the VOCs and the SIPOs can be seen together — and the SIPOs usually far outnumber the VOCs.

SIPOs must be doing something right: they are New Zealand's most common native wader, numbering 90,000 birds at last estimate (2010). Before they were protected in 1906, however, their numbers were not looking so great. They were hunted by both Māori and Pākehā, considered a tasty bird for the table. Protection and probably conversion of tussock to pasture (giving them more nesting sites) have helped this bird.

As their common name implies, but their scientific name *unicolor* belies, variable oystercatchers come in different shades — black, pied or intermediate (smudgy). They're not fussy about the colour of their partner, and their numbers flip the further up the country you go. In the deep south most are dark, with 94 per cent of them black and just 5 per cent pied. Towards central New Zealand, about 85 per cent are black and 7 per cent pied, while in the north just 43 per cent are black and the rest are intermediate and pied. The younger birds all have a brownish tinge to their black feathers, which disappears as they get older.

This variation can make it hard to tell the difference between a pied variable oystercatcher and a SIPO. However, they are ever so slightly different: SIPOs are smaller and have a white notch in front of the folded wing, and more of a clear line between their dark wings and neck and white breast. And as mentioned, variable oystercatchers don't migrate: apart from some who join winter flocks with other VOCs or even SIPOs, many tough it out year after year with their faithful partners in the same territory all year round, making their nests (just basic scrapes, like all oystercatchers) on a sandy or shingle beach, especially around the mouths of rivers and estuaries, or on shoreline pastures.

The chicks of the VOC and the Chatham Island oystercatcher have been seen diving and 'flying underwater' to escape perceived threats. While trying to catch them to band them, ecologist John Dowding has described them swimming out onto the water, propelling themselves with their feet, but then diving suddenly to about half a metre under the water, wings beating in shallow beats and feet stretched out behind the body. Birds then resurfaced up to 10 m away. All oystercatcher parents are vigilant in guarding their chicks; as well as attempting to chase away aerial predators such as black-backed gulls and harriers, they'll sneak from their nest and feign a broken wing to lead predators away.

There are about 5000 VOCs and they're long lived: one bird notched up more than 30 years. It's thought the VOC and the SIPO became different species only about 15,000–30,000 years ago — and today, despite their different breeding habits, diet and size, the two species are genetically almost identical. Interbreeding has happened in a few pairs in north Canterbury, but because one species goes inland to breed and the other stays on the coast, there are few opportunities for hybridising.

The Chatham Island oystercatcher is critically endangered. There were 309 birds in 2010, up from just 52 in 1970, and they live mostly on the northern coast of Chatham Island, where are protected

from their main threats: cat, weka, livestock trampling and storm surges. In the south of the group they also have to deal with skuas, so many hide their nests under bushes or overhangs.

Despite some of these threats being relatively recent, CIO have long been rare. The Moriori hunted them when they arrived at the Chathams 500 years ago (oystercatcher bones have been found in middens) and Pākehā records suggest the birds were uncommon by the 1860s — but people were still killing birds for private collectors and museums.

LEFT
The characteristic
territorial display pose.

ABOVE
South Island pied
oystercatchers and
variable oystercatchers
are often seen together
during the winter, foraging
on beaches or roosting in
large groups.

Penguins

Rainforests and burrows: New Zealand has penguins like no other

Yellow-eyed penguin

Māori names	Latin name	New Zealand status	Conservation status
Hoiho, takaraha	*Megadyptes antipodes*	Endemic	Nationally endangered

Little blue penguin

Māori name	Latin name	New Zealand status	Conservation status
Kororā	*Eudyptula minor*	Native	Declining

Snares crested penguin

Māori name	Latin name	New Zealand status	Conservation status
n/a	*Eudyptes robustus*	Endemic	Naturally uncommon

Erect-crested penguin

Māori name	Latin name	New Zealand status	Conservation status
n/a	*Eudyptes sclateri*	Endemic	Declining

Eastern rockhopper penguin

Māori name	Latin name	New Zealand status	Conservation status
n/a	*Eudyptes filholi*	Native	Nationally vulnerable

Fiordland crested penguin

Māori names	Latin name	New Zealand status	Conservation status
Tawaki, tawhaki, pokotiwha	*Eudyptes pachyrhynchus*	Endemic	Nationally vulnerable

Penguins are found all around the southern hemisphere: some brave the icy coasts of Antarctica, the black-footed penguins are found in South Africa and South America, and one species — the Galápagos penguin — even lives in tropical conditions at the equator.

New Zealand has three penguin species on the temperate mainland — the hoiho, tawaki and kororā — plus a whole host of punky crested penguins further south on the subantarctic islands.

Hoiho

Golden eyes and a yellow headband from eye to eye are trademark features of the yellow-eyed penguin or hoiho. Capable of waddling more than a kilometre inland at dusk to nest, and possessing an ear-piercing shriek, it is New Zealand's biggest mainland nesting penguin (it can reach 6 kg), but is also one of the most endangered penguins in the world.

The hoiho sits in pride of place on the New Zealand $5 banknote, next to a clump of Ross lily from the subantarctic — which is where the bird originally came from. It still breeds on the Campbell and Auckland islands, but from about 500 years ago it also started breeding on New Zealand's mainland (as well as Stewart Island/ Rakiura and outlying islands). Before this, the mainland was dominated by the hoiho's cousin, the Waitaha penguin, which was found in small numbers all over the South Island and lower North Island. After early Māori hunted the Waitaha penguin to extinction (the bones have been

Hoiho are subantarctic forest-dwellers, nesting in goblin-forests of southern rata on Auckland Island.

During the night the parents prefer to sleep elsewhere in their territory, away from their huge and demanding chicks.

found in archaeological sites of villages or settlements), the hoiho was able to slowly expand its range northwards from the subantarctic and take over the old Waitaha penguin sites. It's unclear why the hoiho could survive where the Waitaha didn't.

The hoiho is shy and nests away from others, scrambling up to nest in heath or shrubland in the subantarctic islands. On the mainland it'll nest in scrub or flax, or even paddocks that provide shelter such as rocks or vegetation. Hoiho use four major breeding sites on the mainland: Banks Peninsula, North Otago, Otago Peninsula and the Catlins. Hoiho means 'noise shouter' — and its deafening, high-pitched bray is often heard at the nest site when pairs greet each other, accompanied by jerky head movements. Pairs make a small bowl on the ground filled with twigs and grass where the female lays one or two pale bluish-green eggs. When the chicks are a month and a half old, the parents stop brooding them and visit the nests only to bring them food (during the night the parents prefer to sleep elsewhere in their territory, away from their huge and demanding chicks). Where numbers are high, chicks can sometimes form small crèches away from the nest: when a parent returns with food, it will bray loudly for its chick, which will race back to its nest to feed. For the first year, young hoiho have grey eyes and no headband. After about 100 days they can swim out to sea to find small fish, squid, octopus and krill nearby, diving repeatedly to up to about 160 m. The oldest recorded bird lived for more than 20 years.

With fewer than 6000 hoiho in existence, they're probably the most endangered penguin in the world. In 1990 they hit an all-time low of about 150 breeding pairs in the South Island; since then, numbers have been up and down, as high as 600 breeding pairs in 1996 to another low of 255 pairs in 2015. While there's evidence early Māori harvested them and their eggs for food, the historic, widespread clearing of coastal forest for pasture had an enormous impact on their numbers. But starvation, disease, predators, habitat loss and other threats continue today.

Tourism, when not managed properly, is not good for hoiho. Every autumn, hoiho are stranded onshore without food for 25 days while they grow new feathers (they moult all feathers at once, so their coat is not waterproof during this time — usually, the tightly packed feathers are stiff and oily to keep out water, with an air-filled layer of down underneath for warmth). Close encounters with humans stress the moulting birds (this can even damage their new feathers). Being wary of humans can make nesting birds delay their return from sea, too; this leaves chicks vulnerable and hungry. In the sea, natural predators include sharks, seals, sea lions and barracouta.

Diseases can be deadly. In 2004, avian diphtheria, a bacterial infection that leaves lesions in the mouth, killed around half the chicks on Stewart Island/Rakiura and the mainland. Another major chick killer is leucocytozoon, a parasite that affects the blood and organs. There have also been huge, mysterious die-offs of breeding

adults and juveniles in 1990, 1996 and 2013; the culprit was assumed to be a marine biotoxin, but was never confirmed.

Gill nets entangle and drown hoiho when they dive. And because these penguins forage on the sea floor, trawling activities affect their food sources by reducing seabed biodiversity.

Another contributing factor in the decline of the hoiho over the last 30 years is stress on their environment and food sources from climate change, measured by sea surface temperature.

Since the 1980s, the Yellow-eyed Penguin Trust, Department of Conservation, private landowners, scientists and volunteers have all been working hard to increase hoiho numbers, mainly by protecting and enhancing breeding land. All are hoping to beat the odds outlined by a 2017 study which predicted that, without urgent action, the species may have only 25 years left.

Little blue penguin

The world's smallest penguin, this iridescent little torpedo has the important job of keeping some New Zealand bach owners awake at night by shrieking and growling under their floors.

The kororā is a burrow nester, and some of its mainland colonies are very close to humans — such as in Ōamaru,

Wellington and Oban, and on the West Coast, where baches (as well as clumps of dense vegetation) are favoured nesting sites. It uses its webbed and sharp-clawed feet to excavate its burrows, some tunnels twisting and turning for more than 2 m. The holes are easy to see, with the tufts of down, streaks of poo and bits of twiggy nesting material around the entrance. While the penguin is as happy as a clam in these dark, stinking muddy burrows, it's also content in a nest box. In many areas, such as Wellington Harbour and some offshore islands, groups have set up boxes where nesting sites are otherwise in short supply, or as a way of keeping the penguins seaward of highways on the mainland: cars are a major killer of little penguins.

This penguin may be little but it's feisty, especially when defending the small space around its burrow. The males keep up a fearsome reputation in their colonies by sounding a triumph call — a bit like a donkey's bray — after winning a fight. The fights are fierce, punctuated with growls, bashing flippers and yelping, and any penguins eavesdropping on the battle will remember the victory call of the winner: from then on they lie low, heart rates ramping up, whenever they hear the winner in the vicinity of their burrows. This way, the victor can keep his status and his territory without having to fight again and again.

Many people won't be aware of kororā in their area (unless they have them under their floor boards), because the birds come in from a day of fishing at night, when most people have left the beach. They return at nice easy landing zones, preen themselves, then waddle up to their burrows, which can be hundreds of metres inland and even require hops and jumps up steep hills: ornithologist Hugh Robertson once caught one at the 500 m summit of Kapiti Island.

Their chicks are tubby balls with fluffy flippers for the first few weeks, and for their first 10 days or so a parent will be guarding them 24/7 — taking turns with the other parent, who heads out to sea, returning to regurgitate fishy milkshakes to the chicks. But after that, until they fledge at seven to ten weeks, the young are left alone during the day, which is when they are especially vulnerable to stoat attack.

Stoats (as well as ferrets, dogs and cats) are a huge problem, and where there is not predator control the penguins are in decline. This is one reason why kororā populations do a lot better on remote offshore islands than on the mainland. Natural mass die-offs occasionally happen, though: in early 2018, hundreds, potentially thousands, of little blue penguins were found washing up on beaches along the east coast of the upper North Island; it was suspected that La

Penguins of the night, kororā are at risk of being hit by cars where their burrows are cut off from the coast by roads.

Niña weather patterns over the summer affected their food sources. These penguins were starving, and didn't have the energy to fight rough oceans or survive the moult, where they have to stay ashore without food for a couple of weeks over summer.

Disasters like oil spills are also lethal. Kororā were among the birds most affected by the 2011 *Rena* oil spill off the coast of Tauranga; the oiled wildlife facility took on 383 oiled little blue penguins, while 90 were found dead in the wild. At the rehabilitation facility, staff cleaned and rinsed plumage to remove all traces of oil and detergent without damaging the feather structure, before a long process of waiting for the birds to waterproof their feathers by preening. Ninety-five per cent of the cared-for birds recovered and were successfully released back to the wild.

The little blue penguin is often taken to the vet by well-meaning humans during the moult, too. A moulting penguin looks sad, bedraggled and timid, but in reality it's found a place where it feels safe to sit out the period where it cannot go out to sea (with its waterproof feathers gone, it would sink and drown).

Once the bird has its waterproof layer back (penguin feathers have flattened shafts and form a thick layer that insulates

and waterproofs), it forages out at sea. Whirring on through the water using its flippers at bursts of up to 30–40 km/h, it catches mostly fish, but also crustaceans and squid, mainly in the top 5 m of water, although it can dive up to about 70 m. Generally little blue penguins stay within 20 km of their breeding area, but the odd individual may make a long journey — and they have crossed the Tasman.

For a long time it was thought that the fairy penguins of Australia and New Zealand's little blue penguins were the same species, but 2015 research revealed they have quite different DNA. This built on other differences: the feathers are slightly different shades of blue, and they have different 'accents' in their calls. Unexpectedly, the Otago population of little blue penguins is in fact the Australian species — thought to have arrived here just a few hundred years ago.

Crested penguins

New Zealand is the breeding location of four of the world's eight species of crested penguin. These penguins all have yellow crests of feathers above their eyes: relatively tame 'eyebrows' in the Snares crested, punky spikes in the erect-crested penguin and ridiculously overgrown tufts

in the rockhopper. They can fan them out beautifully, displaying them often when courting, and flattening them back into stripes when they hit the water.

Snares crested penguins have the tiniest breeding ground: they breed only on the predator-free Snares Islands, 100 km south of Stewart Island/Rakiura. These islands cover about 300 hectares and are heavily forested with tree daisies. The thousands of penguin pairs come back every spring to land on granite shores swirling with bull kelp. They scrabble up the rocky, muddy steep sides, past the sooty shearwater burrows and Buller's albatross nests, to make their colonies in slippery muddy

hollows or on gnarled horizontal tree trunks. Needless to say, they end up covered in mud. Unusually for a crested penguin, their numbers have been stable — most others are in decline.

The **erect-crested penguin** is the largest of the New Zealand-breeding crested penguins. When dry, its crests are stiff bristles that stick straight up like a punk's hairdo. Its main breeding stronghold is a couple of island groups far to the south-east of New Zealand: the Bounties and the Antipodes. Nesting space is tight on the Bounties, where these birds form packed colonies on great slabs of granite, raucously fighting over space, sometimes

'Sometimes it is sadly hilarious to see crested penguins call out at an egg that rolled 30 cm or so out of the nest in desperation, apparently in the hopes that they could cheer it on to roll back by itself.'

having to deal with fur seals and Salvin's albatrosses doing the same. There is plenty more space on the Antipodes, where erect-crested penguin colonies have declined massively in the last half-century or so.

The female lays a small egg, then a second egg that weighs almost twice as much and is about 20 per cent larger in size (all crested penguins lay these mismatched eggs: the erect-crested penguins have the biggest size difference between the two). The smaller egg is almost always gone by the time they lay the second — it usually rolls out of the nest and is grabbed by skuas. The nests are virtually barren rock, and so with the wag of a tail the egg can be lost. The penguins won't do much about it, either

— as penguin expert Thomas Mattern says, 'Once it is out, it's gone. Penguins do not try to roll their eggs back into the nest like, for example, geese. Sometimes it is sadly hilarious to see crested penguins call out at an egg that rolled 30 cm or so out of the nest in desperation, apparently in the hopes that they could cheer it on to roll back by itself.' Sometimes a skua can pluck the egg from the nest itself, because the penguins don't start incubating their eggs until the second egg is laid — they just stand beside it. Despite this rocky start, the first egg sometimes does make it, and in some crested penguins the parents sometimes raise both chicks. The chicks are guarded for three weeks and then form crèches in the centre of the colonies

Tawaki are found around the coasts of Fiordland and Stewart Island/ Rakiura, but make massive migrations into the subantarctic during the non-breeding season.

for six weeks, before heading out to sea themselves.

Eastern rockhopper penguins are the smallest crested penguins; within the New Zealand region they breed in the subantarctic Campbell, Auckland and Antipodes groups. These penguins display gravity-defying gymnastic prowess: bouncing and scrabbling ashore up kelp and boulders in huge surf, they scale almost vertical rock, taking huge risky leaps between ledges and using their sharp beak and claws to get a grip. However, rockhoppers have experienced huge decline: Campbell Island's population plummeted by 95 per cent between 1942 and 2012. A major reason is thought to be food shortages brought on by rising sea

temperatures: they have to swim further for their food (to productive areas), and have less to feed their chicks. Predation by sea lions and other natural enemies may add to the decline.

The tawaki or **Fiordland crested penguin** is a forest nester, just like the Snares crested penguin. However, it has shunned the subantarctic islands: it raises its chicks in the dense, moist forests or coastal shrublands of the southwestern South Island and Stewart Island/Rakiura, plus nearby offshore islands, out of the way of humans and other mammalian threats. A shy and secretive penguin, the tawaki quietly scuttles up stream beds and into its nests in the densest forest it can find just as night falls. Here it lays two eggs

The erect-crested penguin is one of the most difficult penguin species in the world to see, nesting only on the subantarctic Bounty and Antipodes islands.

in a dark hidey-hole under logs and roots, solo or in small groups of penguins. The first egg is always slightly smaller than the second — and the second egg hatches first. Of the two fluffy dark grey chicks, usually the larger one is raised, but in good years they'll both make it.

Once they're done raising chicks, tawaki have eight to ten weeks to build up a decent layer of fat to last them through the moult, when they replace their entire plumage in one go — which means three weeks without going out to feed. They need to put on about 2 kg and then lose half their body weight in the moult. But strangely, tawaki will swim south to the Subantarctic Front for this food (a round trip of up to nearly 7000 km) even though they have plentiful food on their doorstep! The Subantarctic Front is a productive region of ocean surrounding Antarctica. University of Otago scientists think the incredible swim — which doesn't seem the most logical way of conserving energy — is undertaken by instinct, inherited from ancestors that bred on islands close to the Subantarctic Front, which is where most other crested penguins breed. In 2017, however, one tawaki swam 5000 km to Western Australia — turning up

malnourished and with a foot injury. Named Roxy, she was rehabilitated and sent to Sydney's Taronga Zoo, where three other tawaki are already held (the world's only zoo to hold them).

This penguin of extraordinary endurance got its name from a god in Māori mythology. The early missionary and writer Richard Taylor wrote the following of Tawaki: 'Originally men were not aware that he was a god, until one day he ascended a lofty hill, and someone who was cutting brush wood saw him throw aside his vile garments, and clothe himself with the lightning: they then knew he was a god.' Perhaps its lightning-like crest earned the penguin its name.

It's not known how many tawaki there are left: threats to their existence include set nets and trawl nets, introduced predators, such as stoats and dogs, and curious humans. If people disturb their nests, these shy penguins will run away, leaving their chicks and eggs to die — and if they are forced to flee from their moulting sites, they may use precious energy and starve to death.

Petrels & Shearwaters

Hardy ocean wanderers

Northern giant petrel

Māori name	Latin name	New Zealand status	Conservation status
Pāngurunguru	*Macronectes halli*	Native	Recovering

New Zealand storm petrel

Māori name	Latin name	New Zealand status	Conservation status
n/a	*Fregetta maoriana*	Endemic	Nationally vulnerable

Hutton's shearwater

Māori name	Latin name	New Zealand status	Conservation status
Tītī	*Puffinus huttoni*	Endemic	At risk

Sooty shearwater

Māori names	Latin name	New Zealand status	Conservation status
Tītī, hakekeke	*Puffinus griseus*	Native	At risk

Grey-faced petrel

Māori names	Latin name	New Zealand status	Conservation status
Ōi, tītī	*Pterodroma gouldi*	Endemic	Not threatened

Westland petrel

Māori name	Latin name	New Zealand status	Conservation status
Tāiko	*Procellaria westlandica*	Endemic	Naturally uncommon

Petrels are true hard, salty, seasoned mariners, spending their lives expertly navigating the stormy ocean, turning up on land only to breed — and, like a pirate that only occasionally comes ashore, most are awkward on terra firma right from the touchdown, with many species having to crash-land through trees at night. But, unexpectedly, if there's no launch pad nearby, some can nimbly climb trees and rock faces with their sharp claws and beaks to take off again in the morning.

Like all seabirds, petrels are perfectly equipped for life at sea; they can drink all the seawater they like to no ill effect. This is because they have special glands near their eyes that flush out the excess salt, which drains out through the nostrils. But while gulls and terns have slit nostrils on their beaks, petrels and albatrosses have external tubular nostrils: enormous lumps running part-way down their bills (petrels have just one dorsal tube; albatrosses, two).

These huge nostrils give them an excellent sense of smell to zero in on morsels of food out on the vast ocean surface, and later to find their partners on land, or (in burrow-nesting petrels) to locate their burrow among the millions of other burrows in the dark.

Since petrels are such seafarers, you're unlikely to see them without the use of a boat, despite tens of millions of them breeding in our waters. Classifying petrels can be confusing — there are four families within the tubenose group: petrels (which includes shearwaters, prions, fulmarine petrels and gadfly petrels), storm petrels, diving petrels and albatrosses. The seas around New Zealand host the breeding sites of 38 species (excluding albatrosses). They have tell-tale silhouettes: hooked beak, a rounded body, strong pectoral muscles, long wings and a short tail. Some spend their lives skimming the ocean surface, seizing floating carrion or small fish, and others plunge for food — shearwaters can dive to more than 65 m, wings switching immediately into swim fins. Some surface feeders, such as the white-faced storm petrel, use their legs as much as wings for propulsion — they bounce on two feet along the water's surface as if on pogo sticks, then spread wings out and glide long distances. The giant petrels have a special tendon that locks the wing, so they can sustain soaring for long periods of time — entire oceans, in fact — without tiring.

Like all seabirds, petrels start off life as complete landlubbers. Some, such as the New Zealand storm petrel, grow up in burrows deep in the bush, the young probably considering themselves forest birds, having known no different in the first months of life. Hutton's shearwaters make their burrows high in the Kaikōura ranges, making them the world's only alpine seabird. With their powerful webbed feet and claws, the burrow-nesting species of petrels are fast and furious diggers that can turn cliffs and hillsides into Swiss cheese. When researchers hike some areas of the Chatham Islands, where prions and storm petrels breed very close together, they can easily crash through the topsoil into the vast networks of burrows just below the ground, crushing any occupants. If they simply must get to an area, people occasionally use 'petrel boards' — bits of plywood under their boots — like snowshoes. All burrow-nesting petrels get home to their burrows by crash-landing into the forest canopy or shrubbery at night, after foraging at sea all day. This can be dangerous: some petrels, such as broad-billed prions, have a huge head that can get caught in the forks of branches, fatally hanging them.

The Cape petrel and giant petrel nest not in protected burrows, but on open cliffs. To protect their chicks from predators (such as the fierce skua), and from inquisitive researchers, they can vomit foul-smelling stomach oil, projecting it with surprising accuracy at attackers more than a metre away. The stench of digested calamari or the bright purple of squid ink can be quite the deterrent! Giant petrels not only have great defence but are also extremely aggressive — they can catch other seabirds on the wing then eat them,

Hutton's shearwaters
breed only in the seaward
Kaikōura ranges, but an
artifical colony has been
established at the end of
the Kaikōura Peninsula too.

One of our miracle birds, the New Zealand storm petrel re-appeared in 2003 after being thought extinct for over 100 years.

Grey-faced petrels are true pelagic foragers, feeding well beyond the continental shelf waters.

first battering them to death on the sea surface then tearing them apart with a powerful beak. They've even been seen stabbing surfacing penguins, or holding them underwater until they drown.

Petrels lay just one egg, and it's massive. The tiny storm petrel's egg is equal to 29 per cent of its body weight — even outdoing the kiwi on this score. As a petrel chick grows, it needs more food, of course — so it will have to wait for ever-longer periods while its parents fly further out to sea, to plentiful fishing grounds, returning only every few nights to regurgitate for it.

The chicks of some petrels are a traditional food source to Māori and still part of the identity of generations of certain whānau. Called tītī or muttonbird, the obese chicks of many petrel species breeding around New Zealand were once taken from burrows before they could fly. Plucked, gutted, cooked and preserved in their own fat for winter larder supplies, their fatty, fishy flavour is still a delicacy for many. These days, only sooty shearwaters (from 36 small islands around Stewart Island/Rakiura) and, to a much lesser extent, the grey-faced petrels from islands off north-east New Zealand are taken. Strictly only Rakiura Māori

The tiny storm petrel's egg is equal to 29 per cent of its body weight — even outdoing the kiwi on this score.

in the south and Ngāti Awa/Hauraki in the north have the rights and protocol from generations of harvest to take them sustainably. Muttonbird harvesting is a major commercial and cultural enterprise in the south. The chicks are flightless, easy to find, and extremely fat — in fact, all chicks in the petrel family will, towards the end of their time in the nest, weigh 15-50 per cent more than their parents. This plumping up gives them the energy to shed their down and grow flight feathers, and survive the last period where the parents cut back on the food deliveries. If young petrels are in poor condition they can't produce oil to waterproof their feathers — as soon as they hit the ocean they'd be waterlogged and become shark bait. It's a fine balance, though: if they don't leave their burrows or nests to exercise and shed that baby fat before they fledge, they won't even get off the ground to fly.

After the 2011 Japan tsunami and subsequent meltdown of the Fukushima Daiichi nuclear power plant, there were concerns about the sooty shearwater (which heads to the North Pacific in June-September) eating radioactive krill, squid and fish. It was unknown whether parents would pass on the radioactivity to the chicks during egg formation. Before the 2012 harvest, muttonbirders sent chicks prepared in the traditional way to the National Radiation Laboratory for analysis; fortunately, no traces were found and the harvest could proceed. Interestingly, the antics of sooty shearwaters were what inspired Alfred Hitchcock's 1963 thriller *The Birds*. One day in 1961 in California, thousands of crazed sooty shearwaters were seen regurgitating fish, flying into obstacles and dying on the streets. This was found years later to be nerve and brain damage from eating toxic algae, which had moved up the food chain through the birds' fish prey.

Before introduced predators arrived in New Zealand, many petrels bred on

the mainland, fertilising the forests with their droppings and their dead. Now, only a few species hang on: the Westland petrel, which breeds in the forest around Punakāiki and is so fierce it can ward off many introduced predators; the fairy prions that nest on ledges halfway down Dunedin cliffs; Hutton's shearwater at altitude in Kaikōura; and sooty shearwaters and grey-faced petrels. The rest are confined to offshore islands.

New Zealand storm petrels were believed extinct for many years, but 'came back from the dead', in a story like that of the takahē and the Chatham Island tāiko. They were rediscovered in 2003, having been spotted at sea 108 years after their last sighting. In the interim they were known from only three specimens collected in the nineteenth century — and these were in European museums. Over the next 10 years, teams led by ornithologists Chris Gaskin and Matt Rayner caught and studied more storm petrels, and finally in 2013 birds were tracked to burrows on Little Barrier Island/ Hauturu, where nests and chicks could be found. The search is still on to see if other islands — such as the Three Kings, Moturoa and the Poor Knights — have breeding birds too.

LEFT
Flocks of fairy prions, fluttering shearwaters and Buller's shearwaters gather in the Hauraki Gulf every summer to feed over schools of fish pushing planktonic prey to the surface.

RIGHT
Safe and warm in the Furious Fifties latitude of Campbell Island, a northern giant petrel chick waits for its parents to return with food.

Pūkeko

Mafia mobs in black and blue

Māori name	Latin name	New Zealand status	Conservation status
Pūkeko	*Porphyrio melanotus melanotus*	Native	Not threatened

The blue-suited pūkeko, or 'pooks', are thick as thieves with their family gangs and are ruthless in the face of predators. They'll raise their chicks as a village, and if any stoat, rat or falcon threatens a member of the group, it can be mobbed by a shrieking rabble of family members who peck and karate-kick the offender into submission or even to death.

The pūkeko is a subspecies of the south-west Pacific swamphen — ours arrived from Australia within the last 500 years. When early Māori and, later, Pākehā cleared the forests, the colourful bird thrived in the grassy paddocks and crops, eating seeds, grubs and shoots — the latter it consumes parrot-style, holding a shoot with one foot and using its beak to strip the fibres away and get to the inside. However, pūkeko are happiest living and nesting in swampy wetland, feeding their chicks on protein-rich worms and insects, and sometimes even fish, frogs, rodents and other birds. Pūkeko are often considered pests in farmland and gardens, where they uproot seedlings, pollute water troughs, eat duck and hen eggs, and even rip the heads off ducklings. In fact, they've been raiding crops and pastures since humans first arrived in New Zealand — Māori would have had to chase them from their taro and kūmara, shouting a song that told the wily birds to go away and

Pūkeko are wetland birds, but have expanded their habitat to pasture and parkland since human arrival.

Pūkeko could also be admired for excelling in modern New Zealand where other birds have faltered.

back to their ancestor. Someone who was old and had colourful life experience was said to have become like a pūkeko — kua pūkekotia — and a stubborn, irritating person was said to have pūkeko ears (taringa pākura).

Despite the bad press, pūkeko could also be admired for excelling in modern New Zealand where other birds have faltered: they seem to be everywhere you look, even lurking on the verges of busy urban routes. One of the secrets to their success is their mafia-like family groups, which are fiercely territorial and protective against predators. Pūkeko manage their social groups differently in different parts of the country. Some groups breed in pairs, but many are communal and have four or five dominant birds practising free love. The great thing about having more than a single breeding pair is that they

can defend a huge territory.

Massey University researcher James Dale, working at Tāwharanui Open Sanctuary, north of Auckland, found that a pūkeko's helmet-like frontal shield (an extension of the beak) is the key to its social standing: the bigger that red shield on its forehead, the mightier it is. And if part of the shield is painted black to appear smaller, other birds will act more aggressively and challenge the painted bird, which may then become subordinate. And if that happens, its real shield will also decrease in size in just days.

Shield shenanigans aside, the group is quite the commune. Unusual in the bird kingdom, the dominant male does not carry out mate guarding; instead, he seems quite happy for other breeding males to approach the female he's just mated with. And mating is no romantic affair. The

Afternoon bath, Western Springs, Auckland.

male chases her around, raking her back until she relents and crouches down to let him mate with her. Sometimes, other males will try to mate with her at the same time or just afterwards. This way, she often receives a mixture of sperm, and so who fathered the egg is a mystery. Some matings are incestuous: mother-son pairings are not uncommon in these tight family groups. The birds also have their own rainbow scene, with same-sex matings. The sharing doesn't end there: pūkeko can't recognise their eggs, despite individual eggs looking quite different, and two or three breeding females may lay in the same nest, which is a bowl of grass or wetland plants. With four to six eggs per bird, 18 anonymous eggs can be found in a single nest. This means hatching can spread over more than a week, which is also unusual.

The whole tribe can take turns to incubate — males mostly at night, females during the day. Even young pre-breeders will take a turn on the nest. When a dominant bird turns up to relieve a subordinate from incubation duties, they'll often bring nesting material, which researcher Aileen Sweeney suggests could be a peace offering, designed to minimise chances of a misunderstanding that could damage eggs.

Once the chicks have hatched, the whole tribe helps raise them. Non-breeding birds too will rotate babysitting duties — while the other adults spread out to forage, one

Skirmishes punctuated with harsh screams are a common pastime of pūkeko.

will stay close to the chicks. The chicks are rarely seen, since they nosedive into tufts of vegetation at any hint of trouble, remaining silent and motionless, while the adult will head in the opposite direction hoping to lure the threat away, flicking its tail more and more often if it's agitated. Although the pūkeko usually runs from danger, flapping its wings, it is quite capable of flying, its legs hanging down ungracefully below. It can also swim well, despite lacking webbed feet; even the chicks can swim from day one, sometimes diving to get out of danger.

If attacked, pūkeko are a force to be reckoned with. They can jump in the air to take down a harrier, and when a group of pūkeko descend on a stoat or rat they can rip it apart. It seems that evolving in Australia among snakes and marsupial predators has its benefits.

In New Zealand, licensed hunters may shoot pūkeko in season, and they say it makes a good pie or soup if you cook it right. The feathers are used in fishing flies. In traditional Māori society a pūkeko skin with its indigo feathers would sometimes be rolled into a ball, doused in fragrant oil and worn as a perfume sachet.

Robins
Trusting forest friend

North Island robin

Māori names	Latin name	New Zealand status	Conservation status
Toutouwai, pītoitoi, tariwai, wheko pō	*Petroica longipes*	Endemic	Declining

South Island robin and Stewart Island robin

Māori names	Latin name	New Zealand status	Conservation status
Kakaruwai, kakaruai, toutouwai, wheko pō	*Petroica australis*	Endemic	Declining

Black robin

Māori name	Latin name	New Zealand status	Conservation status
n/a	*Petroica traversi*	Endemic	Critically endangered

Robins are trusting wee birds around humans, fearless since forever. In 1903 conservationist Richard Henry wrote about them hopping up to eat crumbs within an arm's length from him in deepest darkest Fiordland, 'as if they were accustomed to receive them all the days in their lives, though in reality they may never have seen a man before'.

Still to this day, hopping along the ground on two very long, twig-like legs, the males especially can come close enough to eat mealworms out of researchers' hands.

They were named robins for this tameness and their breast patch, but they are not related at all to the true robin found in the UK and Europe — New Zealand robins are part of a group that is otherwise found only in Australasia and New Guinea. New Zealand robins are endemic, much larger than the Australian species and they spend more time on the ground. They're a forest bird, and thrive in mature forest, scrub and even exotic plantations — anywhere there is a canopy with a healthy understorey.

These small birds come in four flavours: North Island robin, South Island and Stewart Island robins, and black robin. The South and Stewart island robins are in fact closely related subspecies. While the first three have a breast patch, their colouring couldn't be more different to the European red robin — instead, they're a

North Island robins are greyer than their southern cousins, with less of a distinct white belly patch.

Walter Buller wrote in 1873 that he had heard Māori say 'ka kanga te manu rā', or 'how that bird swears'.

dull dark grey, with a pale chest patch: all the better to blend into the forest shade. The South Island and Stewart Island robins have more of a yellow-cream belly patch than the North Island's off-white. The black robin is all black and is found only on two protected islands in the Chathams.

Robins may be confiding around humans, but they're a force to be reckoned with among their own kind. Even though they sometimes steal out to feed, drink and bathe, robin pairs defend their territories all year round. There are aggressive disputes at the borders, and even the dawn song can be interpreted as belligerent. Walter Buller wrote in 1873 that he had heard Māori say 'ka kanga te manu rā', or 'how that bird swears'. The robin has a little patch of white feathers above its beak which it displays when angrily defending its nest.

The males are the bullies in the relationship, often getting their hackles up and chasing away their partner to hog the food sources for themselves — mostly insects and sometimes small ripe fruit — although the females will get their own back and sometimes change mates.

To feed (which they spend about 90 per cent of their days doing), New Zealand robins will perch on branches and logs, then dart down to the ground to grab any prey they spot. They also rummage through leaf litter and use the technique of trembling a foot — it's thought the vibrations flush out bugs, which the birds promptly stab. Flicking the wings and tails also helps disturb prey.

Robins also have the strange habit of caching their food. A study on the South Island robin showed that it will usually eat a bit of its prey, then dismember it

Fledglings follow their parents around for some time after leaving the nest, begging for food while learning to forage for themselves.

and store the remainder for later. Often these treats are earthworms, but also wētā, cicadas, slugs, snails and stick insects, to name a few. The bird may put these leftovers in holes and crevices or forks in branches, either in their territory or outside: anywhere close to where they caught it (the female will sometimes move out of sight of the male before she caches her food — and she'll sometimes raid his stash, too). Sometimes, the robin will kill the prey before storing it, slamming it against a hard object, whereas other times it'll put it still writhing into a cache. Robins can remember the location of up to 12 caches — and which cache had the most food. The male does sometimes show his generous side: during breeding season he shares the loot from his caches with his partner, providing her with a snack a few times per hour while she builds the nest or incubates, and he also feeds the chicks once they are a few days old.

Māori named them for their call: pītoitoi for the male's territorial call (which sounds just like that: *pi-toi-toi-toi*), or wheko pō for their habit of being one of the first birds to sing in the dawn chorus, and one of the last to stop singing in the evening (another Māori name is karuwai/toutouwai for their

watery eyes). During the breeding season the males can sing loudly for half an hour with hardly a pause. Interestingly, the more they eat, the more they sing — and belting out these tunes uses up so much energy that, despite the feast, they are back to a regular body weight by the end of the day. This came to light in a 2010 study on South Island robins: it's thought the singing advertises what great mates the males would make because they're so good at finding food.

Although robins weren't a favourite food of early Māori, they were easily caught for the table with snares — all you had to do was scatter worms, bugs and berries and they'd come right up. Māori would also make them into earrings: the skin was hung up to dry and wrapped around a small piece of wood; it was then scented with vegetable oils and placed in a pierced ear.

Pītoitoi were always present deep inside the forest even when other songbirds were not, so an abandoned house in the forest would be called whare pītoitoi: a pitoitoi's house. Elsdon Best wrote in 1908 that hunting parties in Te Urewera would know their trip would be a failure if they heard the pītoitoi singing deep in the forest —

but in other regions, if it was heard on your right it would be a good sign.

The black robin was once the rarest bird in the world — cat and rat predation, and habitat destruction, reduced the population to single figures in the 1970s, with only one fertile female. She was named Old Blue, and now every single black robin is a descendant of her and her partner, Old Yellow. In 1976 and 1977, a team led by Brian Bell moved seven birds from Little Māngere Island to Māngere Island. Conservationist Don Merton then led the team that each year removed Old Blue's first clutch of eggs from her nest and fostered them with tomtits on Rangatira Island. This meant Old Blue could lay two or even three clutches of eggs per season and grow the population. Now, there are around 250 birds on Māngere and Rangatira islands.

Rock wren & Rifleman

Sole survivors

Rock wren

Māori names	Latin name	New Zealand status	Conservation status
Pīwauwau, hurupounamu, mātuhituhi,	*Xenicus gilviventris*	Endemic	Nationally endangered

Rifleman

Māori names	Latin name	New Zealand status	Conservation status
Tītipounamu, kikimutu, kōhurehure	*Acanthisitta chloris*	Endemic	Not threatened

The rock wren and rifleman are part of an ancient lineage — their common ancestor branched off before the evolution of all other perching birds in the world today. They are the sole survivors of seven recent species; three of these were wiped out after Polynesian settlement, probably by the introduced kiore or Polynesian rat, and the other two after European settlement, one by cats and the other by rats.

Rock wren

The rock wren is not a true wren of the family Troglodytidae. It was just so named because it is small with short wings and eats insects. This tough little critter spends its entire life above the treeline around the Southern Alps, hopping amongst rock jumbles and boulder meadows and alpine shrubs, or huddled in boulder crevices during snowy winters.

The two main subpopulations of rock wren, the northern and southern, are split either side of Aoraki/Mount Cook National Park. They were thought to have formed about two million years ago during a

The commonly used Māori name is pīwauwau, which means 'little complaining bird'.

glacial period, when the two groups would have found separate refuges away from the ice. The rock wren was once common in mountain ranges around the Southern Alps, but has declined. In the Murchison Mountains its numbers tumbled by 44 per cent between 1989 and 2009.

Basically a tiny egg-shaped bird on stilts, the rock wren has barely any tail — accordingly it hardly flies at all, apart from a short whirring flit for a few metres along the ground or between boulders. The plumage is green (some birds are brighter than others) and grey-brown, with yellow sides, cream eyebrow lines and a sprinkle of pale dots on the lower back (these are actually pale feather tips). Long toes and claws help the bird cling to rocks. Bizarrely, the rock wren sometimes wears anklets of dead skin. This reptile-like, scaly translucent skin gradually sloughs off and may be a primitive feature of the moult.

The rock wren is always fizzing with action. Constantly bobbing rapidly, it flicks its short round wings almost faster than the human eye can see, then it's off, hopping and running through the boulder fields on a search for insects in shrubs, mosses and cushion plants, or probing litter and rock crevices. It'll also stalk small spiders on the rocks and lichens, and in late summer eat berries and seeds from alpine plants. Rock wrens have been seen storing caught insects in amongst cushion plants for later.

The rock wren gives a very simple tweet — the commonly used Māori name is pīwauwau, which means 'little complaining bird'.

When winter comes, the rock wrens stay in the alpine zone. How they survive the freezing conditions and with little food is

Rock wrens,
like this male,
are alpine
foragers,
expert at
locating tasty
invertebrates
amongst rocks
and moss.

Rock wrens insulate their nest-domes with anything soft they can find, from feathers to rabbit fur.

a mystery, but it is believed they can lower their body temperature and metabolic rate when the air temperature drops, suggesting they may enter torpor while in their rock crevices, to conserve energy.

The nest is purpose built for mountain conditions that can turn to custard in a flash — and these birds don't skimp on insulation! Pairs build them together above the treeline, in cracks and ledges on rock faces, large boulders and steep banks. They excavate the soil to make a cosy cavity, and inside they create an elliptical or spherical nest that's almost completely enclosed apart from a single entrance

hole. The nest walls are 3-4 cm thick, made with tussock grasses and clad in a layer of mosses, lichens and snow tussock shoots. Inside is where it gets super cosy: the birds collect and line the nest with a luxurious number of feathers that they find about the place: these could be from kea, kākā, kiwi, pipit and, historically, kākāpō (so rock wren nests are a bit like a log of bird species in the area). Once, 414 feathers were found in just one nest on the Haast Range. If they find it, they'll also use fur from introduced mammals, such as possum, red deer or chamois, to boost the warmth. All this investment pays

During the breeding season, male rock wrens will defend their territory against interlopers in short chases and scuffles.

off: during incubation and chick rearing, temperatures in the nest average about 33°C while on the outside it's as low as 0.5°C.

Unfortunately, cavity nesters like rock wrens are easy prey for introduced mammals, even at those high altitudes. Their main predator is the stoat, but even mice can eat their tiny eggs. Rats are more sensitive to the cold, but ship rats do make it up there at times.

Rifleman

The ancient rifleman is like a little green ping-pong ball. Officially New Zealand's smallest bird, it's literally thumb-sized, weighing around 6-7 g, which is no more than a heaped teaspoon of sugar. Like the rock wren, its body seems almost spherical without a long tail to balance it. This micro bird bounces up and down tree trunks after insects, darting and scuttling on its long, strong claws, uttering a constant high-pitched *zip* to its mate — a note so high-pitched that older people often can't hear it — and giving quick flicks of its stubby little wings.

The male is bright green, and the female a more brownish green with streaks across her head and breast. These colours, which

[The rifleman gives] a note so high-pitched that older people often can't hear it.

broadly match the uniform of an early colonial New Zealand solider, may have given the bird its name. Likewise, the Māori name tītipounamu could come from the greenstone colouring as well as the rifleman's cheep — literally 'squeaky greenstone'. The male bird is slightly smaller than his mate, weighing about 1 g less.

In other respects, the rifleman looks a lot like its cousin the rock wren, but has a sharp, slightly upturned beak and shorter legs. Not that it often sees the rock wren, which mainly lives above the treeline, as the rifleman lives in the forest — mostly in beech and tall podocarp around the North and South islands.

Paired birds build a spherical nest in a tree hole (which can be tiny), anywhere from near the ground to very high in the treetops. The male kicks off the construction process, and they both line the nest with leaf skeletons, twigs and feathers. The two to five eggs, although tiny at just over a gram apiece, are proportionately huge if you consider the size of the mother, who lays them two days apart. It's mostly the male who feeds the chicks, along with casual helpers that could be siblings from previous broods, or single males that are eyeing up the young female chicks as they feed them — riflemen can breed at nine months. The breeding pair will often start a second nest while still feeding these fledglings. The raising of up to two broods per season is important in a bird that's in decline.

Male rifleman in beech forest, Arthur's Pass.

Male riflemen, like the
one above, have glowing
green plumage, while
females have smart
streaks on their heads
and backs.

Two young riflemen pester
their mother (centre) for a
feed of small insects.

Royal spoonbill

Rock-star wader with fish-finding sensors

Māori name	Latin name	New Zealand status	Conservation status
Kōtuku ngutupapa	*Platalea regia*	Native	Naturally uncommon

In breeding season, an adult royal spoonbill becomes a badass rock star, donning war paint that includes a yellow breast, popping yellow eye shadow, a smear of red on the forehead and, of course, those rakish plumes.

White crest feathers up to 20 cm long look like white dreadlocks when relaxed, but when raised in display they fan out in a royal headdress, a flash of bright pink skin visible underneath. In non-breeding season, it's the opposite. Gone is the white plumage, spikes and make-up, and in its place is a bird with a funny beak and dirty-looking feathers. You won't hear a royal spoonbill much, but when you do (most often around the breeding colonies), it's a strange one: a cross between a growling dog and a clucking hen, with hisses and moans for good measure.

The spoonbill's *pièce de résistance* is, of course, its gigantic bill. Like a long black wooden spoon, this strange appendage is handy for detecting and scooping up fish, invertebrates, crustaceans and even frogs invisible from the surface. The bird holds its bill slightly ajar and waves it through the water in an arc as it wades forwards, like a metal-detecting beachcomber, disturbing the water and sediment to flush out prey. The bill contains many sensors, called papillae, that detect movement — these allow the bird to find food in murky water or even in the dark (they often feed at night if that's when the tide is right). Having detected something, the bird will grab it quick as a flash between its spoon-like mandibles, snap them shut, lift them skyward, and let any edible mouthful slide down into its gullet. It feeds this way in flocks.

The royal spoonbill is found in Australia,

and also turns up in New Guinea and Indonesia. While it had been spotted for some time in New Zealand (early Māori knew it as kōtuku ngutupapa), it didn't start breeding here until 1949, when it shacked up near Ōkārito's white heron colony and never left, making twiggy nests up high in the exposed treetops of kahikatea. Breeding here gave it 'self-introduced native' status. Since then, yet more birds have arrived from Australia, and now breeding colonies are found scattered around New Zealand, though mainly in the South Island. Numbers have skyrocketed to more than 2000 birds, and thankfully they don't compete with other native birds. Their nests can be found in all manner of places, from near herons high in those kahikatea to down on the ground near estuaries and rivers, rocky headlands, among reeds or in trees over the water, and often near where shags and gulls are breeding. Royal spoonbills will often use their leaf-lined nest of sticks year after year.

At two to three years of age, birds congregate at these breeding colonies in October to find a partner. Unpaired birds

Royal spoonbills are often found foraging or roosting together.

It didn't start breeding here until 1949, when it shacked up near Ōkārito's white heron colony and never left.

have singles parties in trees nearby where they can show off to each other, and do a bit of mock-fighting by jabbing beaks and flapping wings. When courting, they take to the air and fly in circles or figure-eights, followed by some bowing, bill-clapping, headdress-raising and, finally, affectionate mutual preening. When they're greeting their partners, they gape their huge beaks and nod, fanning again those majestic plumes. Pairs stay together throughout the breeding season, sharing the incubation and chick feeding (by regurgitation). Parents keep their two or three chicks by

their sides foraging until all the birds leave the breeding colony for the winter. In New Zealand, royal spoonbills usually head for coastal rivers and northern harbours and estuaries — unlike their Australian counterparts, which tend to favour inland swamps, flooded pasture, rivers and pools.

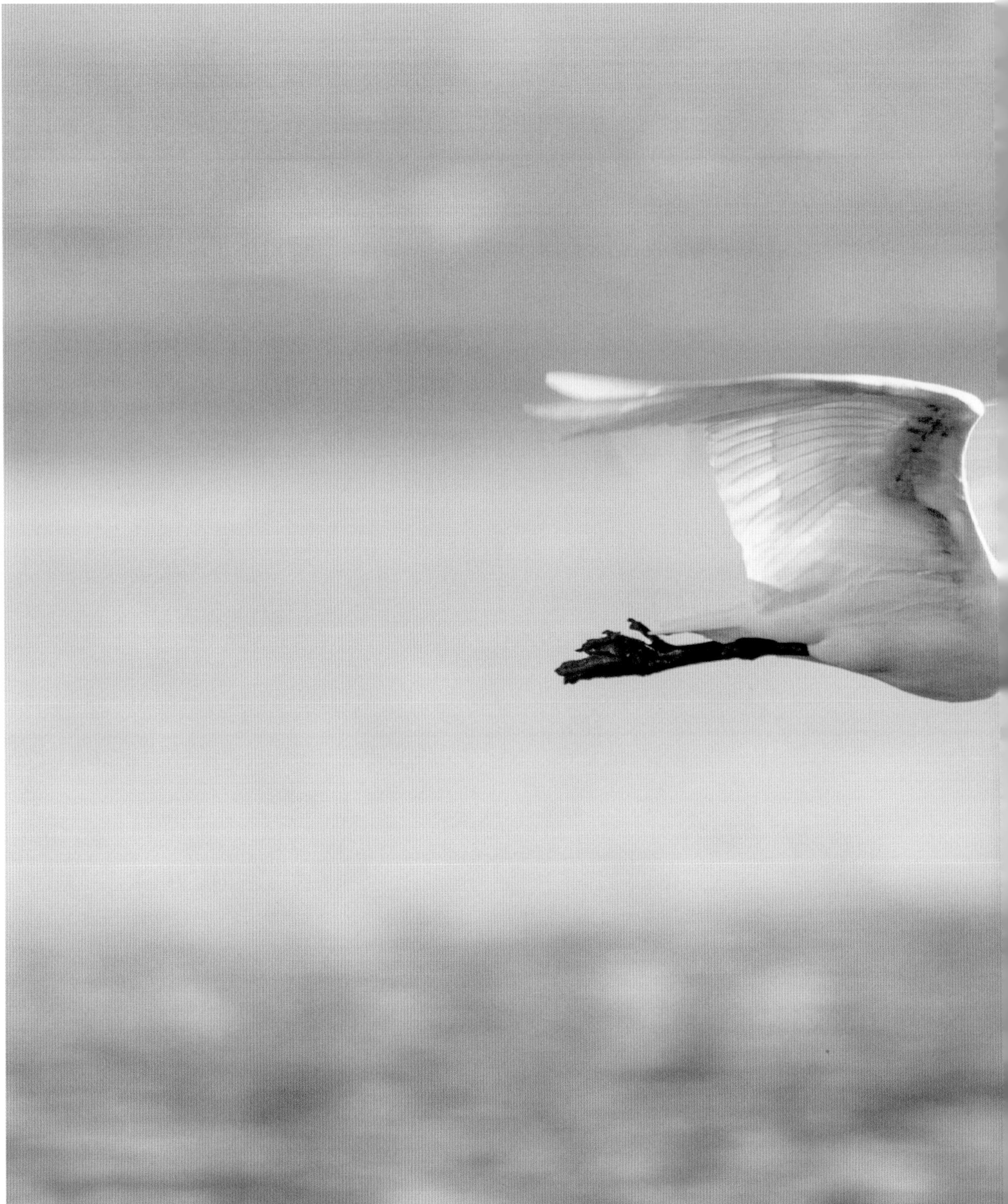

With the
light behind
them, the
wings of royal
spoonbills
are nearly
translucent.

Saddleback

Chattering chiseller

North Island saddleback

Māori names	Latin name	New Zealand status	Conservation status
Tīeke, pūrourou, tīraueke	*Philesturnus rufusater*	Endemic	Recovering

South Island saddleback

Māori names	Latin name	New Zealand status	Conservation status
Tīeke, pūrourou, tīraueke	*Philesturnus carunculatus*	Endemic	Recovering

The saddleback's hallmark call rings out through the forest: a loud, piercing and repetitive chatter that sounds like a car turning over with a flat battery. Follow the racket to its source and you'll see little black birds, each marked as if some tiny being has strapped a chestnut saddle on its back and aims to ride it through the forest. The saddleback's sharp beak is slightly down curved and incredibly strong, and the bird uses it to chisel into bark, which it levers off by opening its beak, looking for all manner of tasty wētā and beetle larvae.

The chattering is the all-purpose call, the contact call. But saddlebacks or tīeke also have quiet fluting calls that males and females use for pair bonding — this is them at their most melodious. Lastly, there's the male's rhythmic songs — the territorial soundbites. These have introductory chips and then repeated phrases that he sings and sings and sings.

The endemic saddleback is a member of the wattlebird family; the red fleshy lobes hanging from the sides of its beak match the blue set of its cousin the kōkako, but its body is only about a third of the size.

There are two species: the North Island bird has a faint line of gold across the 'pommel' of its 'saddle' that the South Island bird doesn't have, and the young South Island saddleback is unique because it is a chocolate brown for the first year of life — the young were once thought of as a different species and were called 'jackbirds'.

Both species of tīeke were wiped out on both main islands by introduced predators in the early 1900s. Later, both were saved by translocations between offshore islands — they would never fly between distant islands themselves, tending to avoid vast open spaces. The North Island saddleback at one point existed only on Hen Island off Whangārei; from these few birds, which were translocated from island to island to mainland sanctuary, today's 10,000-strong population of North Island saddleback is descended. Thus every bird is a cousin of every other bird. And in the south, ship rats raided the tīeke's stronghold of the Big South Cape Islands near Stewart Island/ Rakiura in the 1960s, when the Wildlife Branch (forerunner of DOC) had only just developed the expertise to transfer

The song of the saddleback is a feature of pest-free islands.

saddlebacks between islands. They could save the tīeke, but not Stead's bush wren or the Stewart Island snipe, which sadly both became extinct: death by rat. The South Island saddleback now numbers around 2000 birds. In both species, their various alarm calls, chatters and warbles come in a range of dialects in different populations. This is mainly because they've had to invent their own songs when moved to a new area without any other birds to copy. However, with every translocation and subsequent translocation, saddlebacks have also lost certain sounds in their calls. The original Hen Island population has a wide variety of calls, including gorgeous high-pitched songs, which have been lost in populations down the line.

Tīeke usually nest a couple of metres off the ground in tree holes, cracks in the rock, pungas and epiphytes — when conditions are good, they really get busy and can raise up to four broods per season. The chicks will often stay with the parents for a whole year, foraging as a family group, until the next breeding season: the parents then kick them out of their territory, which is when they wander for three to five kilometres to set up their own territory. Saddlebacks can breed at just one year of age — and live sometimes more than 17 years.

The diet is mostly invertebrates, but a saddleback also goes for fruits and nectar from plants such as flax. It rummages around in the leaf litter, or uses that pointy beak to chisel away up in the trees, raining down debris. If it finds a large insect, it'll hold the prey with one foot while tearing it apart with the beak. Fantails and whiteheads are often nearby: they hear the tīeke calling and noisily chiselling and pecking away, and zero in, staying about half a metre behind and below the larger bird — probably so that they can see the outline of disturbed flying insects against the sky and go for them.

According to Māori legend, the tīeke got its markings from the demigod Māui, who was thirsty after he had snared the sun. He asked the saddleback to bring him water but he was ignored, and so Māui seized the bird, singing its feathers with his hot hand (hence the chestnut markings), before throwing it into the water. For this reason, the tieke was often mentioned in prayers for rain.

Shags

Champions in air and water

Little black shag

Māori name	Latin name	New Zealand status	Conservation status
Kawau tūī	*Phalacrocorax sulcirostris*	Native	Naturally uncommon

Campbell Island shag

Māori name	Latin name	New Zealand status	Conservation status
n/a	*Leucocarbo campbelli*	Endemic	Naturally uncommon

Little shag

Māori names	Latin name	New Zealand status	Conservation status
Kawaupaka, teoteo, pohotea	*Phalacrocorax melanoleucos*	Native	Not threatened

Spotted shag

Māori names	Latin name	New Zealand status	Conservation status
Kawau pāteketeke, kawau tikitiki	*Stictocarbo punctatus*	Endemic	Not threatened

Pied shag

Māori names	Latin name	New Zealand status	Conservation status
Kāruhiruhi, aroarotea	*Phalacrocorax varius*	Native	Recovering

New Zealand king shag

Māori name	Latin name	New Zealand status	Conservation status
Kawau pāteketeke	*Leucocarbo carunculatus*	Endemic	Nationally endangered

Pitt Island shag

Māori name	Latin name	New Zealand status	Conservation status
n/a	*Stictocarbo featherstoni*	Endemic	Nationally critical

Shags are aquatic machines. They can hunt underwater for minutes at a time, swimming incredibly fast like fish-homing missiles with large webbed feet, snagging fish and crustaceans with their quick-fire necks and sharply hooked beaks. The specially constructed feathers are water permeable, which reduces buoyancy and makes diving easy.

But there's no such thing as a free lunch: because the feathers aren't waterproof, shags have to spend time perched afterwards with their wings droopily spread to dry out, before they get too cold (although shags in cold southerly waters don't do this, probably because it would make them even colder).

A shag can swallow a live fish that is enormous in proportion to its beak. Bobbing back up to the surface with the fish, it tosses back its head; the sides of the beak stretch to let the fish slip through, and the huge fish-shaped lump can be seen sliding down the throat. In an ancient tradition in China, Japan and Korea, which in some parts is still practised today, fishermen would raise shag chicks (related to New Zealand's black shag), and train them to dive and catch fish at a command or the blow of a whistle. They'd first attach a ring around the bird's neck so that it couldn't fully swallow the fish, forcing it to wait until it was untied and given

Campbell Island shags are restricted to subantarctic Campbell Island. The Auckland and Bounty island groups also have their own endemic shag species.

a reward at the end. Some shags, such as little black or Campbell Island shags, can hunt together in packs: they dive simultaneously, and group together again when they surface.

Sadly, the diving habits of shags can doom them to be caught in craypots or fishing nets — especially abandoned nets — and drown. Fishing line can also do damage: there is a long-standing colony in the Panmure Basin in Auckland that for the last decade at least has been in decline: overfishing by humans has led to food shortages, and so the shags go for fishers' bait, and end up hooked. The fisher cuts the line, but the trailing line can entangle the shags in trees, leaving them to die.

Shags were eaten in the past. In Mercury Bay and Queen Charlotte Sound, James Cook and his men feasted on shags, roasting, stewing and broiling them, calling them an 'excellent meal'. In traditional Māori culture, adults were sometimes snared on rigged perches, but it was much easier to take young birds from shaggeries (nests) in trees or on cliffs before they could fly. But shags have long been admired by Māori for their strong, direct flight: someone who is travelling resolutely is 'me he kawau ka tuku ki roto i te aro maunga', which means 'like a shag making for a mountain face'. In a haka, forming a straight line in combat is 'kawau mārō', a shag stretched out — the position it takes in flight. Sometimes, shags have been a bad omen. Shortly before chief

In the breeding season, the spotted shag develops a mane of fluffy white filoplumes, and its eye-ring turns a beautiful turquoise blue.

Mananui Te Heu Heu Tūkino II was killed in a landslide in 1846, shags had been seen landing one by one on a rock in Lake Taupō, which was thought to have been an ominous sign. The shag also stars in myth: it was said to be the true identity of Houmea, a woman with an insatiable hunger who used to swallow her husband's entire catch of fish. Eventually, greedy Houmea swallowed her two sons and was later killed, and took the form of a shag.

In the settler era, shags (especially black shags) were persecuted in New Zealand for years and many colonies were completely wiped out. From 1890 to 1940, acclimatisation societies paid bounties for killing shags (again mostly the black shag) because it was thought they were eating too many of the expensively introduced trout. Today, they are protected, although black and little shags can be culled in the vicinity of trout and salmon hatchery ponds.

There are 36 shag species around the globe (most are known elsewhere as cormorants); of the 12 found here in New Zealand, nine are endemic. They come in a staggering array of colours — from the dandified black crest and bright green-blue facial skin of the breeding spotted shag, to the yellow-blue-purple face and white underparts of the pied shag, or the orange facial caruncles and pink feet of the king shag. In fact, shags are grouped according to the colour of their feet. The black-footed species are the ones you see in inland waters, and the pink- or yellow-footed birds are oceanic and salt water only.

The yellow-footed group (the spotted shag and the Pitt Island shag of the Chathams) are unique to New Zealand. They nest in colonies on rocks jutting up from the sea or on cliffs. Chicks of the spotted shags sometimes lose their food to red-billed gulls, which can roost nearby. A pilfering gull will hover on updraughts near the nest, and then, just as an adult shag is feeding its chick, the gull screams loudly, making the parent lunge at it and spill some food — which the gull then nabs. One of the largest colonies of spotted shags is on the Tata Islands in Golden Bay; in winter, thousands of shags fly daily into Tata Beach about half an hour before sunrise to make one hell of a racket: flapping their wings and croaking, splashing in the shallows, swallowing small gizzard stones and regurgitating them back up. At sunrise they leave, flying off in groups to hunt for the day. No one knows exactly why they congregate in such large numbers.

Today, numbers of this colony and others are dwindling. In their northernmost range in the Hauraki Gulf, spotted shags have declined from 10,000 or more in the early twentieth century to just hundreds today, largely on account of

set nets. They can still be found nesting on Tarahiki Island (aptly known as Shag Island) and a few other rocks off Waiheke Island. There is a project under way, driven by ecologist Tim Lovegrove, to attract spotted shags back to the Noises Islands north of Motutapu by 3D-printing shags to create a fake colony, complete with a sound system and white mock poo.

The pink-footed shags are a hardy bunch: lovers of the colder waters, they nest on rocks and dine out at sea, some wrangling octopus and squid and in certain places defending their nests from fearsome skuas. Pink-footed shags include the king shag, Foveaux shag, Chatham Island shag, the three shag species of the subantarctic Auckland, Bounty and Campbell island groups, and, as of late, the Otago shag. It was only in 2016 that the shags on the Otago Peninsula were classed as a distinct species; the northernmost colony at Ōamaru Harbour on the 200 m long abandoned historic Sumpter Wharf is

FAR LEFT
Pied shags roost and nest in trees, favouring coastal pōhutukawa.

LEFT
Whero Rock, off Stewart Island/ Rakiura, is home to a large colony of Foveaux shags.

RIGHT
Both Otago and Foveaux (or Stewart Island) shags have two colour variants: pied morphs and bronze morphs (pictured), which have glossy iridescent black plumage.

now the largest, with more than 300 nests. These shags dive as deep as 50 m.

The black-footed shags have four species in New Zealand. Most of them nest in large trees often overhanging the water, the tufts of nests giving the trees a Seuss-like look; they also nest on rock ledges. Identifying these shags can be tricky. For example, the little shag may be found on salt or fresh water feeding on fish, kōura and even frogs. Its underparts vary from almost all black to completely white, or even variegated ('smudgy'). The juveniles are either all black or white-breasted, and some even have an orange breast from iron stains. To the untrained eye, then, little shags can be mistaken for black shags, pied shags, and little black shags (as if the names themselves weren't confusing enough).

Shining cuckoo

All about au pairs and winters abroad

Māori names	Latin name	New Zealand status	Conservation status
Pīpīwharauroa, wharauroa, nakonako	*Chrysococcyx lucidus*	Native	Not threatened

The world's smallest cuckoo has life figured out. The shining cuckoo swans around the tropics for winter, while an au pair, in the form of the grey warbler, raises its children. This carefree wanderer wears a flamboyant outfit — green-and-white striped belly feathers paired with a shining coat of green iridescence on its wings, back, tail and crown. Amazingly, the plumage blends so well into foliage that the bird is seen a lot less than it's heard.

The sparrow-sized bird spends its winters flitting around the Solomon Islands and eastern New Guinea. It then uses its superpowers of navigation to return to New Zealand in springtime, flying thousands of kilometres back to the area it was born — making it the most southerly ranging cuckoo in the world.

Being a brood parasite, the shining cuckoo owes its existence over millennia to its long-suffering host and diligent foster parent, the grey warbler, and as such its suburban or forest territory usually covers those of four or five grey warblers. A shining cuckoo never builds its own nest nor raises its own young — ever, leaving this work to the grey warbler (and the related Chatham Island warbler). Luckily for the warblers, many — especially in the South Island and lower North Island — are usually on their second clutch by the time the shining cuckoo interferes.

The unwitting grey warbler will go about its nest-building activity as the female cuckoo loiters nearby, watching every move. When the warbler has laid quite a few eggs, the cuckoo will sneak into the pendulous nest while the warbler is away. It removes a warbler egg in its beak, and in its place leaves its own bigger, olive-coloured egg. The grey warbler doesn't realise the egg is different (it's quite dark inside that nest), or doesn't care, and incubates all the eggs together. The cuckoo egg usually hatches first, giving the alien chick a head-start: from about four days old, it begins using its head to shunt the other chicks or eggs from the nest so that it is the sole beneficiary of the food from its tiny foster parents, who can look quite absurd feeding such an enormous chick. The chick has such a convincing hunger call that even other (non-warbler) birds have been compelled to feed it.

The shining cuckoo parent, meanwhile, has no parental duties, but hangs around New Zealand for the rest of the summer anyway, laying down a huge amount of fat as fuel for the trip back to the tropics. Its favourite food is caterpillars — often the toxic or hairy kind that other birds avoid, such as those of the magpie moth, kōwhai moth and monarch butterfly. The toxic hairs become embedded in a thick mucus lining in the bird's gizzard, and then regurgitated.

Because the shining cuckoo is here only for the spring and summer, its trademark whistling call is one of the characteristic sounds of spring: the repeated upward slurs that then trend downward. While many of our shorebirds seasonally migrate overseas, the only forest birds to do so are the shining cuckoo and long-tailed cuckoo. At summer's end the shining cuckoo may gather in loose flocks in preparation for the grand migration.

What did early Māori think of this showy traveller, which they named pīpīwharauroa, 'little bird of the long journey'? According to Elsdon Best, they likened the cry to *kūī, kūī, kūī*, followed by *whitiwhiti ora* — the first call coming in early spring before the fruits of the earth appear, and the second appearing in December, when the food supplies are assured.

Silvereye

The working-bee team, cleaning up insect pests

Māori names	Latin name	New Zealand status	Conservation status
Tauhou, pihipihi, hiraka	*Zosterops lateralis*	Native	Not threatened

Silvereyes or waxeyes are a bunch of particularly hard-working Australians that were probably blown over to New Zealand in foul weather, sometime between the 1830s and 1850s.

Sporting silver eyeliner and olive-green, grey and cream uniforms, forming pairs or vast winter flocks, they descend on trees and crops and vacuum up insects, almost leapfrogging each other as they work an area, staying in touch with their peeping calls. Their nickname of 'blight bird' came from their love for the woolly aphid, which can be a pest, but they love a smorgasbord of invertebrates, including scale insects, flies, caterpillars, mealybugs and their eggs, and spiders.

Their varied diet may partly explain how they've come to be one of New Zealand's most abundant and widespread birds. As well as invertebrates, they're up for pecking at fruits including grapes, cherries and apricots, which means many orchards and vineyards around the country must put up bird nets to avoid serious crop damage, possibly undoing the goodwill bestowed on silvereyes for eating the aphids. Most of their fruit diet is from native trees and shrubs, however — they'll swallow smaller fruits (in the 7–10 mm range) whole, which means they're great at spreading the seeds of natives such as kohekohe, kahikatea, coprosma and mingimingi.

Silvereyes are also partial to nectar: this habit is helped by having a bristly tongue. Since many nectar-feeding birds — stitchbirds, kākā, kākāriki and saddlebacks — are now all relatively rare over most of the mainland, the main nectar-feeding birds that pollinate flowers are the tūī, bellbird and silvereye. The silvereye dines

Other silvereyes are found throughout Asia and the South Pacific islands, including Vanuatu and New Caledonia.

out on the nectar of kōwhai, pōhutukawa, rātā, pūriri and rewarewa flowers, among others. In winter, silvereyes will also gorge on handouts; you can attract them to the garden with mesh bags packed with lard or scraps of fat from the butcher.

While silvereyes tend to flock in winter, in breeding season they split into pairs and establish a territory, defending it with wing-fluttering and aggressive chases, or lowering their wings in a dominant pose. These couples, who bond season after season, weave a cup-shaped nest from mosses, twigs, rootlets and hair, and suspend it among twigs and foliage; in it the female lays three to five pale blue eggs. Both parents share the incubation and feeding duties, raising two or three broods in a season. Considering silvereyes can breed at nine months, and the oldest silvereye recorded in New Zealand lived

11.5 years, their potential to proliferate is strong.

Other silvereyes are found throughout Asia and the South Pacific islands, including Vanuatu and New Caledonia. The New Zealand subspecies also breeds in Tasmania (which may have been where the original migrants came from), migrating north to Queensland in the winter. It is found in the Chathams, subantarctic and Norfolk Island, too. Silvereyes occur widely throughout New Zealand, from the coasts to the treeline, but there are not many seen in deep forest, and their distribution may be changing still. According to the NZ Garden Bird Survey, the silvereye declined by 43 per cent nationally in gardens from 2007 to 2017 — it's unclear why, but warmer winters may mean silvereyes don't come into gardens to search for food.

Silvereyes are social birds, foraging together in flocks and preening each other to reinforce social bonds.

When the bird first arrived in New Zealand, Māori called it tauhou, 'stranger' — and it soon became a valuable source of food. They took the birds using a pae, or a perch, which was a technique also used in the past for stitchbird, saddleback, tomtit and other small native birds when they were more abundant. They used two vertical rods to hold up a perch, and underneath it would be a cord where living decoy birds were tied. The hunter would be waiting beside it in a hide made of branches or fern fronds, shaking leaves and imitating the twittering sound of the tauhou. The curious birds would land on the perch and the hunter would strike them down all at once. The silvereye was easy to catch and plentiful, and Elsdon Best records seeing thousands killed, plucked and cooked whole, preserved in fat, then eaten whole, bones and all.

LEFT
Silvereyes have taken up the job of seed dispersal, which is beneficial to native plants but also encourages the spread of weed species.

RIGHT
Sometimes, foraging for insects requires silvereyes to defy gravity.

South Island takahē

Back from the dead

Māori name	Latin name	New Zealand status	Conservation status
Takahē	*Porphyrio hochstetteri*	Endemic	Nationally vulnerable

One of New Zealand's rarest birds, this eye-catching blue beauty was long thought to be extinct, gone the way of so many ground-nesting endemic birds — and the tale of its 'rediscovery' by an amateur bird fan is a true Kiwi legend. Today, with a dedicated breeding programme under way, the takahē's future is in good hands.

The first recorded European encounter with a takahē was in 1849, when sealers in Dusky Bay followed the bird's footprints in the snow — with the dogs after it, 'it ran with great speed, and upon being captured uttered loud screams, and fought and struggled violently'. The sealers kept it on their boat for a few days before roasting and eating the bird, 'which was declared delicious'. And after a desultory few more sightings, the last being in 1898, this turkey-sized bird was presumed a goner.

Fifty years passed. Then in 1948, Geoffrey Orbell, an Invercargill doctor with an interest in tramping and rare birds, went stag hunting. He had always suspected the bird was still alive, and had worked out from anecdotal sightings where it might be: up on the tops of the Murchison Mountains near Te Anau, in the alpine tussocklands. Lo and behold, he heard the calls, saw the footprints — and later, saw the bird itself: a living takahē. When he returned with the news

In glorious iridescent shades, takahē feathers change colour depending on how the light hits them.

and photos, it made headlines around the world — a species was back from the dead. Away from ferrets, dogs, cats and people, the flightless takahē had lived a secret life high in the alpine tussock. The Murchison Mountains subsequently became off limits to all save some scientists and deer cullers. Later, however, red deer spread to their safe haven and devoured the tussock that the takahē was surviving on, and stoats invaded, too, and at one stage the takahē were thought to number just 118 birds. Now with breeding programmes around New Zealand, the total is up around 300.

The very solid takahē is the world's largest member of the rail family, clocking in at up to 3.5 kg, like a beefed-up version of a pūkeko (also a rail). In fact, in 2015

four takahē were mistakenly shot by contracted hunters in a pūkeko culling operation on Motutapu Island (and in 2008 a Department of Conservation worker shot one on Mana Island). This despite a takahē being three times the size of a pūkeko and flightless. The takahē also has a more colourful plumage of iridescent blue, turquoise and olive green.

Being a rail, the takahē has long taloned toes on powerful legs. When feeding on alpine tussock — the main plant in its Murchison Mountains diet — it snips off the tussock stem with its powerful beak, a bit like wielding a pair of garden secateurs, and holds the tussock leaf in one claw while eating the soft, juicy base. It tosses the rest and keeps feeding — all day in

It all comes out the other end in about eight metres' worth (per day!) of poo.

fact, because there isn't much nutrition in a stem of tussock. So little of the huge amount it eats is digested, it all comes out the other end in about eight metres' worth (per day!) of poo resembling small green sausages or fibrous cigars. It will also use its beak to dig fern roots from the soil. Although herbivorous, the takahē will go for the odd protein source. (In 2011, a takahē was caught on film eating a dead paradise duckling, and a small part of the diet is insects.) Aside from tussock, it snips off the tips of pasture grass, or takes seeds by running its beak along seedhead stems. It'll also use its beak to fight.

Takahē nest on the ground, under cover such as tussock, with their eggs and chicks hidden away from birds of prey — which were the only predators it used to have to worry about before humans and their pests arrived. The female lays one to three buff-coloured eggs, blotched with reddish purple, two days apart. Each egg takes 30 days to hatch; the chick hangs out in the nest for a week, then follows its parents around, mercilessly begging for morsels of food.

The takahē's calls include a loud shriek — think of a squeaking playground swing — and a hooting contact call. It also utters a booming alarm, which one observer likened to someone 'whistling to them over a .303 cartridge shell'.

When breeding programmes first began, the hand-raised chicks would think they were humans, so the Burwood Takahē Breeding Centre, which incubated excess fertile wild eggs, used glove puppets and papier-mâché takahē to brood and feed the chicks to give them more of a takahē identity. As adults these puppet-reared birds weren't as good at breeding, but thankfully there are enough takahē now to use real birds to hatch and rear chicks.

While their native habitat
is primarily tussock land,
takahē do very well on any
grassy pastures.

The North Island takahē (moho) is long extinct, but its southern cousins are thriving in North Island ecosanctuaries such as Tiritiri Matangi.

Spur-winged plover

Masked and armed, but mostly harmless

Māori name	Latin name	New Zealand status	Conservation status
n/a	*Vanellus miles*	Native	Not threatened

The spur-winged plover is a dramatic-looking wader. It hides its face behind outsized wattles, like two pieces of yellow kitchen gloves, which give it a crazed superhero look. Matching yellow eye rings and beak complete the costume — and then the bird reveals the fearsome long, thorny spurs on the carpals ('wrists') of each wing. When it opens its mouth, the shrill, grating call can be heard miles away.

Originally from Australia (where it is called the masked lapwing), the spur-winged plover is now also native to New Zealand, as well as Indonesia, Papua New Guinea and Timor-Leste. It began breeding in Invercargill in the 1930s and never looked back. This black, white and brown immigrant soon found a land perfectly suited to its lifestyle, and by the 1970s it was widespread, from open wetlands to stony and sandy riverbeds and estuaries. Unlike most of New Zealand's endemic birds, it thrives in places with lots of open grassland, such as farms, parks or airports. A key to its success is an unfussy diet: spur-winged plovers will gobble whatever's on hand in the manner of insects, worms, molluscs and crustaceans.

These successful birds used to be protected under the Wildlife Act, but in 2010 their protection was revoked — they were getting on the wrong side of horticulturalists, airports and possibly other native birds. From 1999 to 2004, spur-winged plovers were responsible for 37 per cent of aircraft bird strikes nationally, costing the operators hundreds of thousands of dollars in repairs. Crop farmers were finding damage to cauliflower, broccoli and lettuce, and there is even video evidence of a spur-winged plover destroying a dotterel egg. Our masked superheroes had gone too far.

But how dangerous are they? They often try to look threatening by revealing their spurs (done by flexing and lowering the carpals and raising the head) and inflating their wattles. If the intruder is a human, a harrier or livestock, they'll dive-attack, but they'll rarely make contact. With smaller birds they regard as a threat or competition, like magpies and starlings, they'll run at them and peck them. It's really only other spur-wings that endure full wrath: during the breeding season, pairs set up a territory, with the male

The nest is nothing flash: it's a scrape lined with whatever they can find.

performing a song flight around the boundary. To defend it from another spur-winged plover, they'll noisily call and carry out aerial attacks on the intruder, smashing it on the head with their wings, or charging it with pecks and wing blows.

After mating, the male chooses a nest site and begins flicking material towards it, then pretends to sit in the nest — if the female agrees with his choice, she'll start flicking material into it, too, to make a lining. The nest is nothing flash: it's a scrape lined with whatever they can find, from twigs and stones to sheep poo. They build in wetlands, but also in pasture, where rushes or bits of dead wood can help disguise the nest — even if they risk livestock trampling their eggs — or even on roofs, from which the chicks plummet to the ground unharmed.

Mum and Dad both incubate, taking turns to sit while the other takes sentry duty. If the non-incubating bird spies a predator, it will give an alarm call, making the incubating bird jump up off the nest, run away and start feeding innocently somewhere else, leaving the camouflaged eggs hidden. The chicks can feed themselves almost straight away but are still brooded by the parents for some time. They're little fluff-balls of down with a tiny, colourless mask just starting to form from the corners of the eyes, and they're super camouflaged for most settings. The parents are fiercely protective — dive-bombing, swooping, screaming with their grating call or stumbling away from the chicks faking a broken wing. The chicks develop their trademark spurs at around week three, though they're just tiny grey spikes. The spurs are eventually bony but have a keratin tip, which is shed and renewed every year; the rumour that they are venom-tipped is, thankfully, an urban myth. The spur-winged plover seems fierce, but it's more bark than bite.

Stilts
Wading ballerinas

Black stilt

Māori names	Latin name	New Zealand status	Conservation status
Kakī, tūarahia	*Himantopus novaezelandiae*	Endemic	Nationally critical

Pied stilt

Māori names	Latin name	New Zealand status	Conservation status
Poaka, tōrea, turuturu pourewa	*Himantopus himantopus*	Native	Not threatened

Black stilt

The black stilt or kakī lives up to its name — everything about it is long and thin. It has an elongated neck, a long thin bill and the legs of a supermodel, which trail out behind it when it flies. A red eye pops from jet-black plumage. The kakī also happens to be one of the world's most endangered wading birds.

The kakī may be endangered, but it's hardcore: while almost all other birds head away for winter, most kakī tough it out in the bone-chillingly cold Upper Waitaki Basin in the Mackenzie Country, wading the braided river channels. Here, most winter days sit below 5°C (and can reach -30), shrouded in thick fog (while summer has some of the South Island's driest, most searing weather, with regular nor'west winds of over 100 km/h). But being alone in winter means the kakī gets the unfrozen river deltas, swamps and tarns all to itself. Wading through the shallows with a frosted back and iced-up pink legs, it stabs its long beak at aquatic insects, larvae and small fish, and upturns small stones. It even feeds at night, where it scythes its beak sideways, sweeping through silt to filter out small worms.

The kakī is aggressive, too — it doesn't like strangers. In captivity it will reach a point of irritation where it will even drown other kakī that are unfamiliar. It's also curious: kakī will swarm around any new object they find.

Once, before the advent of introduced predators and environmental change, the kakī was widespread throughout the South Island and central-eastern North Island. But there's now only around 130 adults left in the wild. Eggs, chicks and in some cases adults are wiped out by cats, stoats, ferrets, weasels and hedgehogs — especially because kakī nest on river and stream banks, which are predator highways. When all you can do is feign a broken wing to draw predators away from your nest, you're essentially defenceless.

The vulnerable chicks look like long-legged balls of fluff. As ornithologist Bob Stidolph bluntly wrote in the *Evening Post* in 1939 of the pied and black stilts: 'Contrary to the helpless condition and somewhat repulsive appearance of passerine birds when hatched, the stilt chick and those of other members of the wader family are decidedly attractive and pretty little creatures.'

Wearing pale feathers spotted with

The kakī may be endangered, but it's hardcore.

black, the chicks are almost invisible among the river stones where they hide when falcons fly above them or their parents make alarm calls. Frozen motionless, they're easy to step on, and also easy prey for cats and other hunters. Chicks are exhausted after the hard work of hatching, and will sleep for most of the first 24 hours, but then they're up and at it. They can forage for themselves in the shallow water, but still need their parents to keep them warm. Sometimes you'll see an adult standing with six pairs of legs — two downy chicks tucked up inside its feathers. It's almost impossible to tell male and female kakī apart; researchers have to analyse their DNA to find out.

Black stilts and pied stilts utter a soft clucking when nest building, but generally are known for their high pitched *yep* call. To the trained ear there is a slight difference between the two species. Bob Stidolph likened the call to another species of animal altogether: 'The stilt has a characteristic cry which closely resembles the yapping of a puppy. The resemblance is in fact so great that on one occasion the writer was completely mistaken. He had patiently stalked a supposed stilt and found that it was a Pomeranian.'

Today's population has been bred up from a low of 23 birds in 1981 by the Department of Conservation's Kakī Recovery Programme and the Isaac Conservation and Wildlife Trust, who sometimes use trained dogs to find eggs on impossibly camouflaged nests. They then incubate and hand-rear the chicks in a facility in Twizel, releasing the juveniles after winter in August-September. Adding a new aviary in 2017 has enabled them to hatch and rear even more kakī chicks

The gravel riverbeds of the braided Tasman River are home to most of the wild kakī population.

Post-mating displays for kakī involve fluffed-up feathers and bulging red eyes, as pairs strut side by side.

before releasing them into the wild; mimicking a braided river, the aviary has river pebbles, flowing water and tussocks.

The amazing thing about kakī is that, despite their scarcity, they are very accessible to public viewing. Unlike the kākāpō, which is confined to off-limits islands, you can drive up the road in the Tasman Valley and walk onto the riverbed, and chances are you'll see our rarest wading bird.

Pied stilt

The ancestors of kakī and poaka, or pied stilt, came here from Australia, doing so centuries apart; the first birds to establish themselves had already evolved into kakī, the only black stilt in the world, by the time the poaka arrived in the early nineteenth century.

The poaka is the black-and-white version of the kakī, and is a lot more successful in the modern New Zealand environment — they're found all over the country. The kakī will drive away poaka entering in their territory, but they can't compete with sheer numbers. Poaka number around 20,000–30,000 and don't mind what kind of wetland they inhabit: brackish or fresh water, swamps or South Island braided rivers. And being such a recent arrival (they're also found in the Philippines, Indonesia and Australia), they are wilier than kakī when it comes to mammalian predators: they brood on non-existent nests to distract the predator away from the real nest; they gather in colonies for

Pied stilts are vocal in defence of their territories, yapping and chattering while dive-bombing interlopers.

protection; and they will readily dive-bomb intruders.

Poaka also escape the cold of winter, leaving inland breeding sites and heading to warmer northerly coastal areas offering better food resources. Hauraki Māori, encountering the bird in winter, admired its dignified air and long, purposeful strides, likening the important chief Pāoa to the poaka or turuturu pourewa:

Ka kōhure a Pāoa, me te turuturu-pourewa te āhua e haere atu ana.

Pāoa was taller than any of them, walking along like a poaka.

But kakī and poaka are still so closely related that they will interbreed. The hybrids come in all sorts of combinations of black and white and were once thought to represent about ten different species! Kakī naturally prefer other kakī; but when their population crashed in the mid-twentieth century, leaving them little choice of mates, they began breeding with poaka. Whether kakī hybridise or inbreed, the outcome is never good — either there are fewer eggs, they don't hatch, or chicks don't survive long. DOC's Kakī Recovery Programme is helping to offset inbreeding by boosting the kakī population with hand-reared birds and, on the recommendation of geneticists, sometimes culling hybrid stilts that pair with a kakī, if there are other suitable kakī mates in the area.

Stitchbird

At home in the swinging sixties

Māori names	Latin name	New Zealand status	Conservation status
Hihi, kōhihi, kōtihe	*Notiomystis cincta*	Endemic	Nationally vulnerable

The male hihi is a bright flash of gold on black in the forest, flitting from branch to branch with uptilted tail, white ear tufts and a yellow breastband — he's showy all right, which is completely appropriate for a bird with such an outrageous sex life.

Promiscuity and infidelity? Read on! These small birds couple up in pairs, but in many cases both male and female will be having their way with other birds at the same time, and so 60 per cent of a male hihi's nestlings may not be his own.

To keep up with this lifestyle, a hihi's testicles are four times bigger than they should be for a bird of its size, and productive to match. What's more, the hihi's cloaca engorges, changes angle, and acts like a penile erection (in the bird kingdom, external organs for mating are usually reserved for waterfowl). This 'penis' also enables male hihi to wrestle females into mating face to face, making

hihi the only birds to have sex this way. This 'missionary position' mating is never consented to by the poor female, and usually happens on the ground when she is fertile and just about to lay. She will have put on weight, which is perhaps what the male looks for (her existing partner also guards her more closely at this time).

Though it's tough on the lady hihi, there may be an advantage to all this extra-marital sex. In pre-European times hihi were common, but with introduced predators (and perhaps bird diseases) they have become one of New Zealand's rarest birds, with only a few thousand in existence. Inbreeding is a real threat to a

Many hihi in ecosanctuaries have unique colour-band combinations, which help to identify them.

This swinging habit of the birds keeps the population far more genetically diverse.

healthy population — but this swinging habit of the birds keeps the population far more genetically diverse: with more partners, more birds can pass on their genes, and more compatible egg and sperm matches can take place.

As aggressive as they are in the mating department, hihi are the underdog of New Zealand's three main nectar feeders. They're chased off by tūī and bellbirds, and often have to wait until these two have had their fill of nectar, fruit or insects before they can feed, making lean years even leaner for hihi. Strangely, they're not even related to these two competing species — DNA analysis reveals they have their own family, Notiomystidae, which is closer to wattlebirds (saddlebacks and kōkako) than to the tūī and bellbird. Hihi were long considered to be in the same honeyeater family as the tūī and bellbird,

and the recent revelation that they are not related may explain their anomalous habit (for a honeyeater) of nesting in tree cavities. Saddlebacks are cavity nesters, but the larger kōkako are not.

Though stitchbirds were once common all around the North Island, by the 1880s cats, rats and possibly also disease had exiled them to Little Barrier Island/ Hauturu, making the island the only place in the world they could be found for almost a century. In the 1980s some hihi were shifted to other offshore islands, and later to some mainland sanctuaries, including Wellington's Zealandia. Hihi are susceptible to fungal infections, possibly due to the stress brought on by sexual aggression and competition for food. This competition is worse at translocation sites compared to the untouched, diverse forest on Hauturu,

which has never been impacted by introduced possums, goats or deer.

Why were they named stitchbird? Early New Zealand naturalist Walter Buller thought one of the hihi's calls sounded like the word 'stitch'; others think it comes from the clicking call which sounds like a machine stitching, and yet others say it sounds like two river rocks being struck together.

Hihi have unusually large, bright black eyes and strange rodent-like facial whiskers around the short beak. The female is more understated than her partner — she's an olive grey or brown with the same white wing bar as the male and just a hint of white on each side of her head. Her alternative name in Māori is matakiore — 'rat-face'. To early Māori, the male's bright shoulder feathers were valuable for decorating cloaks, adding a spark of colour. The hihi and saddleback, or tīeke, share a mythological connection. The story goes that the demigod Māui created the hihi's colourful feathers: like the tīeke, the hihi refused to fetch water for Māui after he had slowed down the sun, so in a fit of rage he threw the male stitchbird into a fire, and its feathers became the colour of the flames.

LEFT
Male hihi have white 'ears'
— little tufts of feathers
they can raise at will.

ABOVE
High-pitched hihi calls
can sound like machine
stitching.

Subantarctic snipe

Small insect-eater with a chilling nocturnal alter ego

Māori names	Latin name	New Zealand status	Conservation status
Tutukiwi, hōkioi, hākuai, hākuwai	*Coenocorypha aucklandica*	Endemic	Naturally uncommon

The snipe is a small insectivorous bird that looks a little like a miniature kiwi — and one of its Māori names is tutukiwi — but this fragile, innocent-looking creature has an alter ego: under the cover of night, it fills the sky with an eerie roaring sound that has been compared to a jet engine passing overhead.

The source of this sound was the stuff of legends for hundreds of years — to Māori the sound was made by a mythical bird of prey called hākuwai or hōkioi, which lived in the heavens and visited earth only at night, as large as a moa and colourful. To hear its frightening sound was a bad omen and could mean war. It was heard especially on the Tītī Islands, when muttonbirders were harvesting.

The fearsome sound is actually of the snipe's nocturnal aerial displays, now officially referred to as hakawai displaying. First, from high in the air comes a call of *queeyoo queeyoo queeyoo* or *hakwai hakwai hakwai*, and then *whoosh* — the hair-raising roar of air whistling through feathers as the bird plummets invisibly towards the ground in the darkness before climbing up into the air again.

Intriguingly, no one has seen them do it: the only evidence in the daylight of our snipe's night-time antics is extremely worn tail feathers, with snapped tips, from the stress caused by producing this strange noise. Related snipe of the genus *Gallinago*, found around the world, are known to make a similar sound with their modified outer tail feathers, and they have been seen in their courtship display flights at dusk and in the moonlight.

Ornithologist Colin Miskelly made the link between hakawai displays and snipe tail wear after recording the noise in the Chatham Islands in the 1980s. He also noted that accounts of hearing the hakawai on the Tītī Islands (where the South Island snipe once existed) had died out with the introduction of predators — and then in the 2000s he caught Snares Island snipe with frayed tail feathers, too. Since then, he's found frayed feathers on subantarctic snipe, and hakawai has been heard on four of the subantarctic islands.

Snipe are barely known to most New Zealanders, despite the fact they used to be everywhere, and it's the same tired old story as told for most of our birds: introduced predators wiped them out. The last North Island snipe was seen on Little Barrier Island/Hauturu in 1870, and the last South Island snipe in 1964 in Taukihepa/Big South Cape Island, near Stewart Island/Rakiura. The snipe on the main islands of the Chathams, Auckland and Campbell islands went the same way. Now, they're confined to just a handful of remote offshore islands.

Snipe species are found worldwide — they all have a long, skinny bill and cryptic plumage. The term sniper comes from snipe hunting in the northern hemisphere; the common snipe in Europe is a game bird with such an erratic and fast flight that hunters had to hone their sharp-shooting skills to bag it.

New Zealand has the most primitive of all snipe: the genus *Coenocorypha*. And there are only three species left: the subantarctic snipe, the Snares Island snipe and the Chatham Island snipe. Before human contact, there were nine species spread throughout Fiji, New Caledonia, Norfolk Island, mainland New Zealand and the subantarctic islands, but the usual predators — especially rats — have decimated numbers and driven six species to extinction.

Although snipe are categorised as waders, they don't really wade — rather, they hang out in the dense forest, tall tussock and cold-climate herbfields of their remote islands. They nest in leaf litter or a loose cup of leaves under cover, the female laying two large mottled eggs. Each parent looks after one chick each.

Most active in the morning and at night, a snipe sticks its beak into the soil, making little holes all over the place, to find invertebrates such as earthworms, beetles and spiders, as well as insect larvae and pupae. It pauses in mid-probe, detecting any movement, and can swallow smaller prey even while its beak is still buried.

The subantarctic snipe is the least known of the lot. About the size of a blackbird, it has a longer beak than other snipe, and it's morphed into three subspecies on its respective island groups: the Auckland Island, Antipodes Island and Campbell Island snipe. They're all slightly different: for example, there's a yellow shade to the feathers of the Antipodes Island snipe and a dark back and pinkish belly in the Campbell Island version, while the Auckland Island snipe can be all sizes and have a rusty rump. While these snipe seldom fly, unless disturbed by people, they have evidently flown between islands, presumably at night.

The Campbell Island snipe is a story of beating the odds. It was wiped off the island when rats arrived from a sealer's ship in the early nineteenth century, before naturalists even knew that the bird existed. But in 1997, almost 200 years later, some snipe were miraculously discovered on the nearby tiny, craggy and inhospitable (rat-free) Jacquemart Island. In 2001, the rats were eradicated from Campbell Island, and by 2005 snipe had flown back and were finally breeding in their ancestral home. Fittingly, the Campbell Island snipe was given the *perseverance* subspecies name by Miskelly.

The Auckland Island snipe was once spread throughout its island group, but visiting ships and failed attempts at settlement had brought pigs, cats, rabbits, mice and livestock to many of the islands — eventually, only Adams Island, Disappointment Island and Ewing Island were free of these invaders and could sustain snipe. By 1993, however, mammals were cleared from Enderby, Rose and Ocean islands, and the snipe naturally flew over and moved back in. But strangely, each island has a different breeding 'clock' — at any one time there will be eggs on one, juveniles on another, and a mixture on the third — and no one yet knows why.

The Antipodes Island snipe is present throughout the Antipodes Islands; the only introduced mammal that established there was a mouse, arriving sometime in the nineteenth century. Curiously, this snipe pauses breeding for two and a half months in summer, while the other snipe are busy at it — it's suspected the mice play a part, gobbling up all the invertebrates that the snipe need to sustain egg-laying and chick-rearing. In 2016 the 'Million Dollar Mouse' eradication project was carried out to exterminate mice, and scientists are interested to see the effect on the timing of breeding in snipe in the future.

Elusive and shy, subantarctic snipe are at home in dense tussock.

Swamp harrier

Roadkill ripper on rural highways

Māori names	Latin name	New Zealand status	Conservation status
Kāhu, kērangi	*Circus approximans*	Native	Not threatened

The kāhu is one seriously fierce-looking vacuum cleaner. It soars and glides lazily through New Zealand skies, scanning the ground for dead animals. Dead sheep, roadkill possums and hedgehogs, uncollected hunting kills: whatever the carcass, the tawny kāhu is the bird for the job.

Sometimes, it'll want something a little fresher, in which case no small being is safe: birds, rabbits and hares, eggs, rats, frogs, fish, lizards and even large insects are all fair game — the kāhu's slow flight or hover suddenly shifts into a steep dive ending with wings spread, sharp crushing talons that knead and pierce, and a gory ripping open of soft flesh (like the eyes and mouth) with a hooked beak. The harrier's extra-long legs help it scoop up prey in long vegetation. If they detect it approaching, other birds will flee their nests and chicks jump to their deaths on the ground rather than face this big killer, which usually snatches chicks from the nest and takes them elsewhere to consume. With the female weighing about 850 g (and the male 650 g), the kāhu is not to be trifled with.

Also called swamp harrier, Australasian harrier or harrier hawk, the kāhu breeds in parts of New Guinea, Australia, southern Pacific Islands and New Zealand. In Australia, it tends to stick to swamplands — perhaps because in that country it faces competition from the 23 other species of raptor. In New Zealand it feeds in open

Their high-pitched whistles are heard from far away as the male does his 'sky dance'.

country, such as wetlands, farmland, tussockland, scrub, mangroves and riverbeds. The only places you won't find it is in densely forested ranges like Fiordland and in urban areas. In fact, it has been so successful (thanks to the open country created when European settlers cleared forests and introduced a population of tasty rabbits), that at one point acclimatisation societies accused it of attacking lambs and cast sheep, and eating game-bird chicks. They put a price on its head which remained in place until the 1940s. The bounty led to the slaughter of hundreds of thousands of birds, but even this didn't make a dent in the population until rabbit control in the 1950s began limiting their food source. Kāhu have been partially protected since 1986: if individual harriers are harming property or are a risk to endangered birds, they can be killed.

Kāhu are silent most of the year until the breeding season, when couples return to the same territory they hold each year. Then, their high-pitched whistles can be heard from far away as the male does his 'sky dance' — making steep dives again and again from high in the sky. Sometimes the female will flip over in a barrel-roll to claw him mid-flight, and sometimes they'll briefly lock talons. The male will spend a lot of time marking his territory, soaring round with wings held in a steep 'V', long legs hanging down. Kāhu are usually monogamous, although the occasional male will have two mates, making him a very busy bird. Nests are platforms made with all sorts of sticks, stalks and rushes, usually in swamps, or on bracken, blackberry and other low bushes — the females are fickle incubators and will desert the nest if disturbed by humans.

Kāhu have good eyesight and a piercing gaze.

The male feeds the female and the chicks by passing food to his upside-down mate while they're both in flight near the nest — something they will have practised in their courtship flights. The chicks hatch over a few days, and when older than two weeks, they'll fight each other for food, striking out with their talons. It's really a case of survival of the strongest: the first chicks that hatch grow so fast and claim all the food, that the younger chicks die — but their carcasses don't go to waste: they become food for their older siblings. Chicks are brave from the get-go: if there are intruders, the chicks will throw themselves on their backs, hiss and strike with both razor-taloned feet.

At the end of summer, kāhu leave their breeding territories and fly to a spot up to 100 km away for the winter, sometimes roosting each night in communal flocks (more than 100 birds in previous times, when they were more abundant) in swampy areas, hunting alone during the day. The oldest kāhu recorded in New Zealand was more than 18 years old — and like humans who go grey, the kāhu's feathers pale into old age.

These birds figure in early Māori lore. If a tribe was meeting to decide whether to go to war, to see a kāhu flying above the village signified that it was the right choice. Often, a meeting of chiefs was named 'e hui o ngā kāhu' — a gathering of kāhu. The feathers were used as head adornments (piki kāhu) and to decorate items, such as wooden weapons, and when early Māori were making kites (for pleasure and for divining) they sometimes made a huge harrier bird kite — manu kāhu — in honour of the demigod Māui, who would often fly in the shape of the kāhu.

Kāhu were sometimes snared in those days with a long, pronged rod with bait and noose at the end. However, some tribes would never eat the bird because they considered it a messenger between humans and the gods, others because it was a carrion eater that probably fed on human corpses. And considering the sight of kāhu standing on roadkill is familiar on rural roads, they were, chillingly, probably right.

Kāhu are extremely wary of people, immediately soaring away if they feel observed.

Terns

Pure grace on pointed wings

Black-fronted tern

Māori names	Latin name	New Zealand status	Conservation status
Tara pirohe, tara piroe	*Chlidonias albostriatus*	Endemic	Nationally endangered

Fairy tern

Māori names	Latin name	New Zealand status	Conservation status
Tara iti, tara teo, tara teoteo	*Sternula nereis davisae*	Native	Nationally critical

These small and slender birds are poetry in aerial motion, but with an alter ego of absolute ferocity. Despite their size, they'll take on anyone who nears their nest like a bat out of hell, dive-bombing, dive-pooping, chattering with fury. Of the 37 or so tern species world-wide, six breed in New Zealand and another eight visit regularly. Two of our terns are endemic: one breeds on braided rivers and the other is the smallest and rarest of the lot.

Black-fronted tern

The black-fronted tern is endemic to New Zealand — nowhere else will you find this endangered wee bird with its smart black cap and white cheek stripe. In the summer, out on the braided riverbeds of the South Island, it breeds in loose colonies on little gravel islands in the river (it's the one tern to breed only inland). In winter it becomes coastal, roosting in hundreds-strong flocks in harbours and estuaries, and this is when you'll sometimes see it up in the North Island, from Wellington to Hawke's Bay and sometimes even the Kaipara Harbour.

To see these terns foraging buoyantly

Any predator threats and the adults will sound the alarm and sometimes dive-bomb as a group.

on the wind is like watching an aerial dance, especially along the braided rivers and grassy flats, where they pinpoint small insects on the ground and swoop down to pick them off. All sorts of food is up for grabs as these gourmands move between their summer and winter homes. Also known as 'ploughboys', they'll gobble flying insects and juicy earthworms from pasture or ploughed fields, with the birds working in flocks, snapping bugs from mid-air or plucking them from the ground. Mayflies and small fish are on the menu when the tern hunts alone over braided river channels, working its way upstream, dipping its beak or plunging right in, then circling back downstream and starting all over again. It also snaps up skinks from river flats. In winter, it takes crustaceans in estuaries and lagoons, or near the surf zone — and one black-fronted tern was even seen feeding 35 km out to sea.

Back on the braided rivers, black-fronted terns nest in the same area every year. First, pairs will carry out flirty aerobatics, such as synchronised gliding, and exchange food gifts, such as fish, skinks or earthworms. Their nest is a shallow bowl on sand or between rocks, where they lay one to four eggs and take turns incubating them. The chicks don't venture far: they hide and wait for their parents to bring them food until they are able to fly and hunt for themselves. Any predator threats and the adults will sound the alarm and sometimes dive-bomb as a group. Early Māori ate the eggs of tara pirohe, harvesting them annually from the islets in the braided rivers. Ethnologist Herries Beattie was told by a 'Maori matron', 'The eggs were lying about all over the place, and it was easy to go round with a basket

Early spring snows are no problem for black-fronted terns, who forage on without regard for sub-zero temperatures and snow-bound riverbanks.

and pick them up . . . the eggs were small and dainty . . .'

These eggs are not so plentiful today. Having evolved to nest in a braided river system, the black-fronted tern is caught in a bind, now that its breeding habitat is disappearing and predators have moved in. Braided rivers are gravelly networks of channels that are always shifting and changing, washing away new vegetation and creating bare little islands with a protective water 'moat' around them: this constantly renewing gravel is where the bird evolved to forage and nest. Introduced weeds, such as willow, lupin and broom, stabilise the shingle islands, making straighter channels rather than the meandering, renewing braids — and the resulting vegetation creates places where introduced predators can hide. When water levels drop because of drought, hydroelectricity or irrigation schemes, this increases vegetation growth and so the space available for breeding also drops.

But it's the introduced predators that are the main reasons why these terns are endangered. The adults, chicks and eggs have no chance against a lethal

Like all terns, black-
fronted terns are fierce
in defence of their nests
and young.

combination of feral cats, ferrets, stoats and hedgehogs. Just one predator can wipe out the nests of a whole colony. In some areas, however, the kāhu (on the lowland Wairau River) or black-backed gull (on the lower Waitaki River) is the main predator — and even pied oystercatchers have been caught taking eggs. There are thought to be fewer than 10,000 black-fronted terns left.

Fairy tern

The fairy tern, with its smart breeding outfit of black cap, orange beak and panda-bear eyes, has the frightening title of the most endangered native New Zealand bird.

The fairy tern's Māori name is tara iti, 'small tern', and it is the smallest of the terns that breed in New Zealand. It's related to the Australian and New Caledonian fairy terns. There are about 45 birds left, and only about nine breeding pairs (up from a low of three breeding pairs in 1984). Once, fairy terns were found on many beaches around the North Island and the east coast of the South Island. Now there are just five sites where they breed — Waipū, Mangawhai, Te Ārai,

Pākiri and Papakanui Spit. The birds are heavily managed in the Department of Conservation's fairy tern recovery programme: eggs are transferred between nests to minimise inbreeding, nesting areas are fenced off and trapped for predators, rangers are employed to look solely after fairy terns, and Auckland Zoo provides facilities for incubating eggs that are abandoned or at risk of being swamped in a storm.

Part of the problem for fairy terns is how they nest, in a New Zealand that has changed a lot in the last couple of centuries. Between October and February, these birds scrape a small hollow in the sand to lay eggs among the seashells. Unlike most other terns, this one prefers to nest alone rather than in colonies, and so it doesn't have the advantage of being able to mob predators en masse. Its bumblebee-sized eggs and chicks are superbly camouflaged, and the 360-degree clear view around the nest provides an unhindered line of sight to approaching danger. Still, introduced mammalian predators will track fairy tern nests by smell, and they can get closer to the adults than ever before because of the introduced dune plants that provide cover. These dune plants also take away nesting habitat. From the 1930s, European marram grass and other exotics were introduced systematically by the government to stabilise New Zealand's dunes, because the naturally wide, low and volatile sand dunes were drifting inland onto valuable land: fairy terns of the past would have enjoyed predator-free beaches that stretched hundreds of metres from waves to dunes, with countless nesting spots away from storm surges. Now, beaches are mere slivers of what they once were: they are lined with steep marram grass dunes that cannot be nested in, and the existing sand lies in the path of storm surges.

Another problem is human disturbance. Beach-goers unwittingly scare adults off nests, drive over eggs, bring chick- and egg-hungry dogs, and trample nests with horses. Papakanui Spit, off Kaipara South Head, is a Royal New Zealand Air Force weapons range. Despite this dangerous title, it's the perfect home for fairy terns: out of bounds to humans and their troublesome activities, and even the defence force stays away from the spit during fairy tern breeding season. In fact, in August 2018 air force personnel drove a Unimog with 12 tonnes of shells and sand to the spit to create two large, compacted mounds that it's hoped will protect nests against the high tides and storms. Predator trapping is still needed in all areas, however, to control the host of cats, rats, ferrets, stoats and hedgehogs, which can sniff out the adults, eggs and frozen-when-threatened chicks. The black-backed gull and kāhu are also partial to the odd egg or chick.

Fairy tern populations aren't the fastest growing, either. While pairs will try to nest up to three times in the same season if their nests fail, they'll only raise one brood, putting months of effort into just one or two chicks, feeding them and then teaching them how to fish, which involves almost hovering above the water before plunging for small fish and shrimp.

Once widespread, fairy terns are now restricted to a few Northland beaches.

Tomtit

Scout of the deep forest

Māori names	Latin name	New Zealand status	Conservation status
Miromiro, ngirungiru, kikitori, torotoro	*Petroica macrocephala*	Endemic	Not threatened

In the deep, still forests of mainland New Zealand, when all other birds are silent, you might encounter the tomtit — a tiny, softly flitting bird with an almost disproportionately large head. In the North Island it has the Māori name miromiro: the verb miro means 'to twist, twirl, or move rapidly' — and that's exactly what the tomtit does.

It perches on branches or trunks, carefully scanning the trees and ground, before darting forwards to snatch an insect in its short, whiskery bill. Or, if you're on a far-flung offshore island, such as the Snares, you might find it foraging on the ground in the tussocklands, or snapping at flies in penguin, seal and sea lion colonies. Wherever it forages, the tomtit's incredible eyesight enables it to spot insects 10 m away; it was famed for this among early Māori, sparking the saying 'he karu miromiro' — 'a tomtit eye', used for an observant person.

Indeed, the miromiro has featured in many Māori ceremonies and beliefs. Because it would turn up out of nowhere in the deep forest, it was nicknamed torotoro, 'scout' — it would be the first to visit the water troughs that were set up to snare kererū. Along with the morepork and fantail, a tomtit appearing inside the house was a bad omen. More helpfully, it would scratch around on ground that had

Sometimes tomtits would be captured and used as a love charm to get a woman to desire someone far away.

been trodden on, searching for insects, and warriors could use this to see where their enemy had gone.

Sometimes tomtits would be captured and used as a love charm to cause a woman to desire someone far away; this was called the ātahu ritual. The man — or the tohunga (expert) working with him — would wait until the wind was right, perform a karakia (incantation) and throw the miromiro into the air, whereupon it would ideally fly straight to the woman, wherever she was, alight on her head and put her under a spell: she would then set out to find this man.

Tomtits were also released to bless events and people, such as when a baby was born, a tohunga initiated, kūmara crops planted, or a pā built, and it was a source of food. Its feathers were sometimes placed in the hair of the dead for honour, along with those of the albatross and the now-extinct huia.

Tomtits come in five subspecies: the North Island tomtit (with white underparts), South Island tomtit, Chatham Island tomtit, Auckland Island tomtit (these last three have yellow underparts) and the Snares tomtit, which is completely black. Generally in tomtits the male has black upper feathers and the female is more grey or brown. The Chatham Island tomtit was used in the 1980s to save the critically endangered black robin — it made the perfect foster parent to incubate the robin's eggs and raise her chicks while the robin got on with laying more.

Tomtit pairs can breed from one year old. They keep their territory all year round and defend it fiercely in the breeding season, patrolling the area and singing from perches with a trilling whistle. While

Tomtits can be found in all parts of the forest, flitting along industriously to catch insects.

Hardy tomtits even manage to thrive in the tempestuous subantarctic, breeding on the Auckland Islands.

the male sings loudly and proudly his few short notes, the female sings the same but almost under her breath, very softly. However, if tomtits come face to face with a tomtit who shouldn't be there, they'll raise their crown feathers aggressively; or if it's another species (including humans), most will show the white frontal spot on their forehead. The female makes her cup-shaped nest of twigs, moss and the like in a tree cavity or fork, in a ponga, or among thick vines. Most pairs (apart from the dense populations on outlying islands) raise up to three broods per season, laying up to six if they fail! Prolific layers they

may be, but they don't live very long; usually to a maximum of about three years, although occasionally birds have reached ten or more.

Before Europeans arrived and brought more predators and further land clearance, this bird was absolutely everywhere, from sea level to the subalpine zone. Today it especially is found in beech forests and mānuka or kānuka scrub, but will be happy in exotic forest, too. Occasionally tomtits turn up on farmland and in suburbia, but these are usually just the young ones still searching for somewhere suitable to set up a new territory and find a mate. One

Female tomtits may be more sombre in plumage than males, but they share the characteristic bright orange feet.

mature bird that was translocated from the Hūnua Ranges, south of Auckland, to Tiritiri Matangi, in the Hauraki Gulf, made it all the way back to his territory, completing a journey of at least 56 km, including long stretches over open water. Their populations are stable on the North and South islands, but in the Chathams they haven't survived where there are introduced predators.

Tomtits are related to native robins but are much smaller, their white or cream belly covers a larger area, they have stumpier legs and a white wing bar, and they don't stand as upright as robins. In one bizarre case in 2010, a male Stewart Island robin was seen by biologist Bryce Masuda feeding a nest of South Island tomtit fledglings — even though he already had a nest of his own nearby. Masuda thinks the robin may not have been able to tell the difference between the two nests, as they were both in tōtara trees of a similar height, and the begging sounds of the broods were similar.

The tomtit also looks a bit like a member of the tit family in Europe and Britain, hence the common name, but they are not related at all — the tomtit is endemic to New Zealand.

Male North Island tomtits
have stark black-and-
white plumage.

The all-black Snares
tomtit forages from
twisted tree-daisy forests
all the way down to rocky,
penguin-strewn shores.

Tūī

A song of exquisite smoothness and breaking glass

Māori name	Latin name	New Zealand status	Conservation status
Tūī	*Prosthemadera novaeseelandiae*	Endemic	Not threatened

The tūī, which can serenade you sweetly one minute and sound like a cough mixed with a sneeze the next, is found only in New Zealand and is thriving. Whether you're in the North, South or even subantarctic Auckland islands, you'll find these minstrels in forests, farms (of the more wooded type) and, increasingly, downtown.

Research has shown urban tūī have changed their habits to nest very high in trees or at the ends of branches where cats, rats and dogs cannot reach — this is street-smart adaptation at its finest. By comparison, tūī in predator-free sanctuaries tend to nest all over the tree, even within reach of the ground.

The tūī was the first native New Zealand species to be protected from hunting. Before the 1870s, there would be bundles of dead tūī hanging up for sale in shops, and huge numbers of skins were exported to Europe for ladies' hats. To Māori the wild tūī were considered a delicacy; they were snared, preserved in their own fat and potted into calabashes to be presented to guests at feasts — and James Cook found them pretty tasty, too. The tūī was taken off the hunting list in 1878, and then absolutely protected from 1906.

Early colonists called the tūī the 'parson bird' because of the two tufts of white throat feathers and black body. However,

when the sun hits its plumage, the tūī shows its pop-star iridescence — at different angles the feathers can appear all shades of blues and greens, with bronze on the back. On its neck the tūī has strange feathers called filoplumes — curved and silver grey, they're almost like hair.

Tūī can be feisty warriors, sometimes extremely aggressive to other birds, and males have been known to fight one another to the death — first clashing in the trees, stabbing each other with sharp claws, then falling to the ground, where they continue the scuffle. This aggression, typical of honeyeaters, may stem from the fact that nectar sources are so clumped: flowering trees are to be treasured.

In flight the bird twists, dives and swoops, opening and closing its wings. It can fly silently, but older or more dominant males can 'switch on' a whirring noise by spreading their notched outer wing feathers. They do this mostly when defending a good food source or showing off to a lady tūī.

Although it will eat the occasional stick insect or cicada, the tūī lives for sipping at flowers. It is the larger of New Zealand's

Fluffed-up feathers
and ultrasonic song are
features of territorial
displays.

two native honeyeaters (the other being
the bellbird), with a tongue that is long,
thin and frayed like an old paintbrush.
The bird flicks its tongue into flowers to
soak up nectar — in the process getting
a scattering of pollen stuck to its head
feathers, making it an important pollinator
of some New Zealand trees and other
plants. It can travel at least 30 km a day to
take nectar from trees such as pōhutukawa
and kōwhai. Harakeke (flax) is a favourite:
it has a flower with the same curvature as
the tūī's beak.

There is nothing quite like the tūī's voice.
This extraordinary vocalist not only has
more than 300 songs, but also mimics
everything from other birds to human
speech, cars and phones. Early Māori
would tame tūī and teach them hundreds
of words, including long speeches to
welcome and entertain guests. Until his

death in 2011, a male tūī named Woof
Woof at the Whangārei Native Bird
Recovery Centre would rattle off a stream
of phrases, all in the exact voice of his
handler.

How does the tūī do this? It has a
songbird voicebox (with nine muscles
controlling the tautness of the syrinx,
the equivalent of our larynx), and like
all songbirds it can also regulate airflow
from different air sacs because of the low
position of the syrinx, meaning it can duet
with itself, making two different sounds
almost simultaneously. Some of the tūī's
sounds are too high for the human ear to
hear — it will seem like it's singing, with
mouth open, yet no sound is heard.

Tūī can be promiscuous. Because
females prefer big males, with big throat
plumes, a regular male tūī may sometimes
be duped into caring for chicks that aren't

his. Researcher Sarah Wells found that typically 57 per cent of chicks in nests belong to another male — this is one of the highest rates globally among socially 'monogamous' songbirds (the average is 11 per cent). This could be what has driven up male body size — male tūī are 50 per cent heavier than females, representing the most extreme sexual dimorphism (in terms of size) in monogamous songbirds.

According to researchers, doing this 'extra pair copulation' is a way of ensuring good genes for your chicks even if you are paired with a low-quality male. This could backfire in small, isolated populations, because if a big male is siring half the chicks in the neighbourhood, it could lower genetic diversity. Luckily, tūī are abundant enough for it not to be a problem in most places.

LEFT
Notched primary feathers
can be angled to make a loud
rushing sound during flight.

ABOVE
Tūī and flax (harakeke) are
perfectly matched — the tūī's
beak is curved to match the
shape of the summer blooms
so that it can easily sip nectar.

Weka

Cunning pilferer

Māori name	Latin name	New Zealand status	Conservation status
Weka	*Gallirallus australis*	Endemic	Not threatened

The flightless weka is a cheeky rail found only in New Zealand, and havoc is what it loves — with a glint in its red eye, this brown bird the size of a chicken relishes nicking food from campsites, raiding vegetable gardens for a snack, stealing chook food and eggs, and pilfering anything novel and shiny, fleeing with it in its strong beak, neck stretched out, running for cover where it can inspect it more closely.

If no one's home, it'll even venture into houses, leaving one hell of a mess — it can jump 70 cm up onto tables, and weka poo really can stain a carpet.

These were once common complaints in New Zealand, and indeed the weka's piercing *coo-et* duet at dawn and dusk (easily mistaken for morepork or kiwi) was a common sound. But weka started dying out in most places in the first half of the twentieth century, probably because of the combined pressures of predation by dogs and mustelids, drought, and habitat loss. Its sole strongholds were a stretch between Motu and Ōpōtiki on the North Island's east coast, and scattered parts of the South Island.

Breeding depends on food supply: when

food is plentiful, some weka can breed like rabbits, having multiple clutches throughout the year (up to 14 chicks from successive broods have been recorded). The female lays an average of three eggs into a woven cup of fine grasses hidden in dense vegetation. The male does a lot of the incubation and most of the care of the dark, fluffy chicks. The young will leave their parents' territory when they're three or four months old, and can breed from five months old.

When weka are in their natural environment away from humans, they eat foliage, new shoots, fruits and seeds. They dig through the dirt with their beak to find invertebrates, and on the coast they'll pick through seaweed on the beach for treats like sandhoppers and snails. Weka will also take rats, mice and lizards, as well as eggs of burrowing and ground-nesting birds. Though threatened throughout much of their range, in certain locations they cause problems for other native wildlife, and for this reason they have been removed from 11 islands where they were introduced by humans.

They do, however, play a vital role in

North Island weka are common but shy around the township of Russell in the Bay of Islands.

the ecosystem. A 2018 study showed the only living birds that disperse the purple berries from the hīnau tree in a viable state are kererū, kōkako, weka and brown kiwi; and the weka does the vast majority. It is thought hīnau may have evolved to be dispersed by flightless birds — because the ripe berries dropping to the ground are dispersed more than those still attached in the canopy.

Weka can thrive in a lot of different habitats, from above the treeline to scrubland, wetlands and coasts. They'll readily swim in the sea and in lakes (they've been known to board boats), and can swim a kilometre or more between islands. They are often found on the edge of forests, where they can run for cover if they have to.

Plumage colour and patterning can depend a lot on location. Most weka are brown streaked with black, but they range from chestnut on Stewart Island/Rakiura to black on surrounding islands and in Fiordland. Because of this huge colour variation, it was long thought there were many subspecies, named North Island weka, buff weka, Stewart Island weka and western weka. But it turns out that the plumage variations in the South Island are

not localised to any particular places. A 2017 study led by Steve Trewick of Massey University showed there are really just two groups of weka that are genetically different enough to be called subspecies: one in the North Island (*Gallirallus australis greyi*), and the other in the South Island (*G. a. australis*). The researchers looked at the weka's DNA and also the DNA of their feather lice; each subspecies had its own lineage of parasites. The buff weka introduced to the Chathams is of mixed origins, and the Stewart Island weka's features occur also in the western weka population.

According to ethnologist Herries Beattie, weka taste like roast chicken with a hint of mutton, and historically weka were an important food source for many iwi and European settlers. Some iwi would catch weka using a long rod with a slip-noose at the end. To lure the bird, they waved another stick with a bunch of feathers on it, a small bird (dead or alive), or a leafy branch, while imitating a bird call. Because weka are inquisitive birds they are easy to catch, and a saying about weka questions whether a person will make the same mistake twice:

Makere te weka i te māhanga e
hoki anō?

Will a weka that has escaped the
snare return?

On other occasions spring snares were set
under hīnau trees with ripe fallen berries,
or set snares were put on the birds' paths.
Some Ngāi Tahu people would go inland
in April to set up weka 'camps' for months
at a time, where the men would go out with
dogs and hunt weka, bringing them back
to the camp for the women to cook and
preserve in the birds' own fat. In August
they would head back to their permanent
homes and store the birds in an elevated
storehouse.

Early Europeans called the weka
'woodhen', and soon came to know its
nature — it even foiled James Cook's
plans to have goose thriving in Fiordland.
While charting New Zealand's coastline,
Cook liberated geese at Dusky Sound,
naming the spot Goose Cove. The geese
didn't survive, and the weka is assumed
to have helped their demise (we know
weka were there because Cook wrote they
'eat very well in a pye or fricassee'). Early
New Zealand historian Robert McNab

personally witnessed a weka raiding the
nest of a swan in Dusky Sound, tapping
the egg and sipping on some of the
contents, less than a minute after the swan
left the nest. 'No imported geese,' he later
wrote, 'could successfully contend with
such an ever present foe.'

Further Europeans soon discovered how
delicious they were. Charles Heaphy wrote
about roasting one for Christmas dinner,
his 1842 recipe instructing cooks to catch a
weka at Rotoiti or Cape Foulwind, stuff it
with sage and onion, and then roast it on a
stick, to be served with damper or pancake
on a saucepan lid.

Weka are still hunted and eaten to this
day in the Chathams, where they are not
native, having being introduced there, as
mentioned, from the South Island in 1905
(luckily, too — because the buff-coloured
weka introduced there soon became locally
extinct on the mainland). They are also
hunted on some of the Tītī Islands. On
Chatham Island, with no mustelids to kill
them off, they're almost considered an
introduced pest, and Chatham Islanders
go by the nickname 'Weka' in the way
mainland New Zealanders call themselves
Kiwis. Some 5000 birds can be sustainably
hunted in autumn, when they're fat

Chatham Islanders go by the nickname 'Weka' in the way mainland New Zealanders call themselves Kiwis.

and without dependent chicks. The Department of Conservation culls weka in some areas of the Chathams, including the breeding places of the incredibly rare Chatham Island tāiko.

You need a permit to bring weka meat from the Chathams to mainland New Zealand; this is because weka on the mainland may not be killed, captured or sold without DOC's permission. But that hasn't stopped people debating the idea of farming native animals such as weka to ensure their survival, proposing serving them up on dinner plates like lamb and beef. For instance, Christchurch conservationist Roger Beattie farms

eastern buff weka. Beattie observes the law — he doesn't kill or sell his weka — and he contributes many of his bred chicks to conservation schemes. But he has eaten weka killed by stoats, and he points out that no farmed animals have ever gone extinct. The question, however, remains: if we farmed it for the table, would it still be its wily scallywag self — or would domestication extinguish that spark?

Weka on
Ulva Island
are regular
beachcombers,
tossing wrack
aside to gobble
up amphipods.

White heron

Seen once in a lifetime for good luck

Māori name	Latin name	New Zealand status	Conservation status
Kōtuku	*Ardea modesta*	Native	Nationally critical

This regal-looking white bird with long, slender legs is the biggest heron you'll see in New Zealand — its neck is 1.5 times the length of its body, and the bird can stand to a little over a metre tall, towering over the other herons when walking on estuaries around New Zealand.

When the kōtuku flies, it hunches its neck back into its shoulders, forming a huge kink. Turn over a $2 coin and you'll see it on the back. Elsewhere in the world it is called the great white heron or eastern great egret. It is a common bird in Australia, India, Japan and China, but in New Zealand there have never been more than a couple of hundred.

Kōtuku have long been revered in Māori culture: a sighting even just once in a lifetime is good luck. The saying 'he kōtuku rerenga tahi', which describes it as a bird of a single flight, has been used when an esteemed but rare visitor arrives — such as when Queen Elizabeth II toured New Zealand in 1953-54. And their feathers were extra special. During the breeding season in spring-summer, some herons, including the kōtuku, put on garments that could be described as bridal: their bill turns from yellow to black, facial skin turns turquoise, and they grow a gorgeous veil of hair-like, lacy feathers across the lower back, which they can fan out like a peacock's tail. These are aptly called nuptial plumes. According to Elsdon Best, to early Māori, each of the

Solitary kōtuku fly from all around New Zealand to breed in a tiny area on a lagoon on the West Coast.

feathers — for example, the longer wing feathers and the nuptial plumes — had different names, and the most coveted were reserved for men (women were told their hair would fall out if they wore them, making them a laughing stock). If a man wearing these plumes was eating with others, no women could join the group to eat unless he took off his feathers. The feathers were also used to decorate kites and sometimes chiefly cloaks.

Once they've grown their impressive feathers, solitary kōtuku fly from all around New Zealand to breed in a tiny area on a lagoon on the West Coast of the South Island, near Ōkārito, where they first settled when they presumably arrived windblown from Australia a couple of hundred years ago. (The 150th and 175th anniversaries of the signing of the Treaty of Waitangi in 1990 and 2015 featured a logo of a kōtuku, to signify that, like all New Zealanders, the heron's origins are in other lands.) Shags and spoonbills, too, have set up colonies nearby. In this crammed colony, ringing with harsh croaks, the kōtuku take partners, the males advertising themselves to females, fanning out their nuptial plumes, snapping, and flicking their wing feathers. Once the choice of lover is locked in for the season, they'll preen each other and intertwine their necks, and build twiggy nests in kōwhai, kāmahi and ponga. Surrounded by a wilderness of tall kahikatea and lagoon, they gorge on whitebait and raise their chicks before flying off again for the winter. Kōtuku live up to 20 years — that's a lot of annual visits back to the West Coast.

The plain yellow facial skin and bill of the kōtuku in the off-season undergo a remarkable colour change in the breeding season: the bill turns black and the facial skin turns bright turquoise.

It can clear out entire goldfish ponds if it finds them — and will also eat the odd small bird or lizard.

Early Europeans also appreciated the silky white feathers, and they almost wiped out the breeding colony after finding it in 1865. At that time it numbered about 60 birds, but with fancy feathered hats being all the rage and herons being slaughtered all around the world, people raided the colony here, too, and by 1940 there were only four nests left. The area gained reserve status the following year, and became a wildlife refuge in 1957. Today, the breeding site, the Waitangiroto Nature Reserve, has regular predator control and controlled visits (you can only view the colony via a guided boat tour). The only threats to the birds are storms, or being hit by cars when they fly low while launching into flight. There are thought to be 150-200 kōtuku in New Zealand today.

When it's not growing incredible outfits, the kōtuku spends its winter days wading through shallow waterways, a solitary figure with long toes spread wide in the mud, watching for fish, frogs and invertebrates, which it spears with its long yellow bill. It can clear out entire goldfish ponds if it finds them — and will also eat the odd small bird or lizard. For almost 20 years until 2010, a cheeky kōtuku the locals named Hamish would turn up every winter at the Māpua wharf near Nelson, stealing bait and fish off children fishing. But now and then he would also attack the hapless sparrows eating restaurant crumbs, spearing them, shaking them and swallowing them whole, much to the horror of diners. He's presumed dead now and his statue stands there on a wharf pile — a permanent tribute to one of New Zealand's most elusive yet magnificent species.

Outside the breeding season, kōtuku spread across New Zealand to forage in wetlands and marshes.

Whitehead, Yellowhead & Brown creeper

Three wee tree-crawlers

Whitehead

Māori names	Latin name	New Zealand status	Conservation status
Pōpokotea, tātāeko, hore	*Mohoua albicilla*	Endemic	Declining

Yellowhead

Māori names	Latin name	New Zealand status	Conservation status
Mohua, hihipopokera, mōhuahua	*Mohoua ochrocephala*	Endemic	Recovering

Brown creeper

Māori names	Latin name	New Zealand status	Conservation status
Pīpipi, pipirihika, tītirihika	*Mohoua novaeseelandiae*	Endemic	Not threatened

These three musketeers are the hopping, flitting social flocks of the forest: in autumn and winter they move in crowds of over a dozen birds like swarms of butterflies, often in constant feeding frenzies with other species like kākāriki, fantails and silvereyes.

Hardly ever coming down to the ground when introduced predators are about, they harvest food right up to the tree canopies with loud cheery chirps and acrobatics, even hanging upside down by their strong legs and claws to forage under loose bark and snap up invertebrates — from spiders to moths, beetles and caterpillars — as well as small fruits.

All three birds are endemic to New Zealand, and all of them have the unfortunate honour of being the long-tailed cuckoo's choice of foster parent (much as the grey warbler is the 'chosen one' of the shining cuckoo). They find themselves laboriously, but diligently,

raising a colossal alien chick reared from an egg that was rudely laid in their nest by a cuckoo, displacing their own offspring.

While the three birds are closely related, a few differences have appeared over the course of their evolution. For example, most birds can see ultraviolet light, which is important for foraging, choosing mates and recognising one's eggs and chicks; but while the yellowhead and brown creeper lost this ability, the whitehead has retained it. It's not known why, but scientists think this may have influenced the mates they choose and explain why the three birds are so different in their plumage colours. And speaking of plumage: the tail of the

Whiteheads are the only member of the genus in the North Island, but they are confined to areas with sufficient predator control.

They use the tail as a genius 'third leg' to support them while they hang on to a branch.

whitehead and yellowhead is a ratty, tattered and worn mess, with the central shaft spiking out the end. This is because they use the tail as a genius 'third leg' to support them while they hang on to a branch or trunk with one leg and scratch around with the other.

The **whitehead** or pōpokotea is found in the North Island only, in beech, podocarp and even exotic forest. The male is unmistakable: with off-white head and chest, and glossy black legs, he chatters and sings all day. The early colonists called him 'Joey Whitehead'. Some Māori (such as those in Whanganui) believed the flocks of these ghostly birds moving through the forest canopy were the spirits of the dead. The bird was tapu, considered a messenger between humans and gods. For example, when a baby was born, a tohunga would take a live whitehead and touch it to the baby's head while reciting a karakia, to give the child mana and prestige. Then, once the bird was set free, if thunder was then heard, it meant the gods would bestow those qualities on the child. The whitehead was also released when a new pā was built (as was the tomtit).

Whiteheads are small birds but powerful en masse: they can form armies and drive away larger birds, such as morepork. Extremely social, they build nests high in the forest canopy or in shrubs, and form intergenerational family groups to share the load: their grown-up kids from previous years who haven't yet paired off will help raise new chicks, and will even sometimes bring the nesting mother her food (the female is a bit more understated than the male: duller, with paler legs, she also sings shorter notes).

The **yellowhead** or mohua is effectively

Mohua can often be found in large feeding flocks that scour the forest from canopy to leaf-litter in search of invertebrate prey.

the South Island counterpart of the whitehead. On the mainland, it is found high up in the canopies of beech forests, where predator numbers are lower than in podocarp and broadleaf forests. (On predator-free islands it thrives in podocarp–broadleaf habitats.) It has a bright yellow head and breast, and is slightly larger than the whitehead. Also known as the bush canary, it's sometimes confused with the introduced yellowhammer, but the latter doesn't go into the forests and the former doesn't come out.

Unfortunately, instead of building nests in the canopy like the brown creeper and the whitehead, yellowheads build them in tree holes — which means they can't escape when a predator, such as a stoat or rat, arrives. Yellowheads are in trouble now, apart from a few populations with effective pest control in the South Island and on island sanctuaries.

Some iwi used the mohua to predict the weather — Ngāi Tahu believed that when a flock of mohua rose like a flock of golden butterflies out of the treetops and then fell back into them, this would warn of an approaching storm.

The **brown creeper** or pīpipi — not related to the American and Australian birds of the same name — is the smallest of the three *Mohoua* and is found in the South Island and on Stewart Island/Rakiura. Although the pīpipi goes largely unseen, owing to its brown camouflage and habit of both foraging and nesting high in the trees, it can be heard: the male's whistles and slurs can make him a great songster (the female has shorter notes), and there are regional differences in dialect. While the brown creeper eats mostly invertebrates (hanging upside down a lot to get them) and sometimes fruit, historically when conditions were harsh it would leave the forest for sheep stations, where it could flit around eating the fat from bones and skins of butchered animals.

Brown creepers can live in exotic pine plantations, scrub and native forest, from the coast up to the treeline. They move between the trees in large flocks. They will mob long-tailed cuckoos in spring and summer — it seems strange that they recognise the adult, but they don't know when there is a cuckoo chick in their own nest. Speaking of which, the brown creeper weaves an open cup-shaped nest, and each of the female's three or so eggs are huge, almost one-fifth of her weight.

Brown creepers are curious, social birds, constantly chattering to their flock- mates.

Wrybill

At one with river stones, with a sideways beak

Māori names	Latin name	New Zealand status	Conservation status
Ngutu parore, ngutu pare	*Anarhynchus frontalis*	Endemic	Nationally vulnerable

The wrybill is the only bird in the world with a beak turned sideways — it always points to the right, at an angle of 12–26 degrees. This small plover uses its wacky beak to probe under stones in riverbeds the same way we would use a spoon to explore dessert in a bowl: that is, sideways. In its breeding ground of the greywacke shingle riverbeds in the South Island, the wrybill can flip a stone and expertly fish out insect larvae, such as caddisfly.

Wrybills migrate north in winter, typically heading for the mudflats and harbours of the upper North Island. Hot spots are the Manukau Harbour and the Firth of Thames (85 per cent go to these two), where they are found roosting on one leg in packed flocks at high tide when the mudflats are under water — often they lift into the air en masse to form incredible aerial displays, all birds manoeuvering together. Many roost on factory roofs in South Auckland. In these winter feeding grounds the wrybill snaps up insects from the ground surface, but also tilts its head

and sweeps its bill from left to right to scythe food out of the mud — something that would probably give any other animal severe neck pain.

From August, the wrybill heads south to find a territory spread along Canterbury and Otago braided riverbeds, places where this bird turns almost invisible, despite being roughly twice the size of a sparrow. With its pale grey and white colouring (with a black breeding band on the male), it melts into the background when stationed on the nest. Its two eggs look like oval river stones, its nest just a subtle wee scrape in the gravel, and its chicks are like little grey self-sufficient and hard-to-spot bumblebees that freeze when alarmed.

This bird evolved in an environment where its main predators were hawks and falcons that hunted visually from the air: so the wrybill, barely visible against the riverbed, was once safe as could be. As it did with so many other New Zealand birds, this strategy became a problem when introduced mammals that hunt by smell came along — feral cats, stoats, ferrets and hedgehogs can pick out the stationary chicks, eggs and adults and make a beeline

for them. Hedgehogs are the most serious threat on the braided riverbeds, every night heading out to snaffle up eggs; one study showed they were responsible for two-thirds of wrybill nest raids.

Today, there are thought to be only about 5000 wrybills left — although they are hard to count, thanks to the camouflage. Other reasons for their demise include vehicles driving on riverbeds, jetboat wash, and picnickers and anglers unwittingly scaring the parents off the nest. When water is taken from the river for irrigation, water levels fall and expose the pebble islands in the braided rivers to predators, also allowing exotic weeds like lupins and broom to gain a foothold: these are great places for predators to hide.

The wrybill's breeding range is now mostly in the upper Mackenzie Basin, upper Rangitātā and the Rakaia. Both parents incubate the eggs, and will perform a distraction display to lure predators away from the nest, but with only mixed success: the bird will often return to its nest too soon, when the intruder is still only a few metres away and able to locate the nest easily.

Miranda is
one of the
favoured
hang-outs
for wrybills
during the
non-breeding
season.

Afterword

— *Edin Whitehead*

When I set out to photograph the birds that appear in this book, I called it a 'birdventure' — an adventure centered around birds. But, in truth, I've been on a birdventure my whole life, one that shows no signs of stopping. I don't remember the first time I 'noticed' birds. It's too far back in the muddled memories of childhood for me to disentangle it, because I've always been enthralled by things with wings. Flight fascinates me. The effortless, innate grace of birds in flight is what draws me in with awe and, of course, a little envy.

But there is so much more to birds than their ability to disregard gravity. Spending time with them, you begin to notice their quirks of character, their individuality. There's something in the simple joy of watching birds go about their lives that I find impossible to explain. Their mechanics of mind are so unknowable to us from within our human perspective. Jakob von Uexküll, a German biologist, called this 'umwelt' — the unique sensory experience that each species has of its surrounding world. For example, how do oceanic wanderers circumnavigate the globe, forage on the scattered bounty of the ocean and find their way back to the place where they were born? What I love about science is our striving to explain these processes, in the full knowledge that we may never truly understand them. We'll never know what it's like to unfold a 3 m wingspan and leave the tussocked slopes of Campbell Island behind, having never flown before, and return five to ten years later after circling the globe on the Roaring Forties. Eye to eye with a southern royal albatross, you see a glimmer of this life beyond the realms of what we can ever know.

In spending time with many of the species that call Aotearoa New Zealand home, I have been awed by their resilience, charmed by their character and concerned for their future. There is no substitute for the experiences you have with wild birds in their environment, observing their natural behaviours. It's the crucial element to creating photographs that capture their essence. It can be frustrating at times when you can't find the birds you're looking for or you don't get the images you had planned but, ultimately, it's worth it. You see the birds on their terms, and you come away with experiences (and hopefully photographs) that you never expected. All of the birds pictured in this book are wild, although some are confined to sanctuaries and pest-free islands, the only places they can thrive. There are so many moments from this birdventure that will stay with me for life, the moments where a fierce wild joy has gripped my heart so hard that I've been breathless.

Watching light-mantled sooty albatrosses in courtship flight over Enderby Island, pairs swooping in mirrored movement on stark pointed wings, their two-tone wheeze-and-wail cry piercing the air.

A rock wren, tiny, tinier than I'd ever imagined, popping up just in front of me in the boulder-field of Ōtira Valley, vanishing in an eye-blink with a peep so high pitched I felt it press against my eardrums.

Sudden spring snow in Eglinton Valley, waking up to a thick duvet of white with flakes still falling, and seeing black-fronted terns foraging along the braided riverbanks, dropping from the sky like snowflakes themselves to pluck away insects invisible to me, their orange beaks aflame against the monochrome world.

Wrapped in pre-dawn darkness on the Poor Knights Islands, the howling, braying, cackling chorus of Buller's shearwaters making the air throb as they depart the colony, skittering up trees with remarkable ease and launching into the lightening sky.

Lying wet in the rivermud, pelted with rain, watching dappled young kakī work their way up and downstream, dipping face-first into the Tasman River to pluck out caddisfly larvae, unconscious of the fact that they are the rarest wading bird in the world.

In the dim grey of a Stewart Island/ Rakiura dawn, picking out the dark bulk of a tokoeka in the muttonbird scrub and listening to her step, probe, sneeze gently to puff dirt from her nostrils, and probe again in search of underground delicacies.

An impact, so slight as to be near-imagined, as something lands on me in the pitch black nocturnal forest of Little Barrier Island/Hauturu. A hush of leaf litter as it vanishes into the darkness, and then, through an infrared camera, two glowing orbs for eyes, a sloped forehead ending with a delicate bill and a streaky black-and-white belly. A New Zealand storm petrel — a bird back from the dead — returning to the one island where it is known to breed. It sits briefly and then raises itself on impossibly spindly long legs, lifts its wings and starts to dance.

So many of my encounters have been with seabirds — and that's no coincidence. Aotearoa New Zealand is home to more seabird species than all of our land and freshwater bird species combined. We have a greater diversity of native seabirds than any other country by a very large margin. They are also the most endangered group of birds in the world. The other half of my life as a seabird scientist tackles the challenges of conserving these birds in a changing world, in the hope that we can prevent the accelerating slide towards extinction for some of our most threatened species. Our unique fauna are national treasures, and we must do all we can to protect them.

As much as we've tried, no word or photograph can capture the true brilliance that is the living bird. To watch them as they go about their lives is to glimpse into another world. So when you put this book down, I urge you to look out the window. Go outside. Birds are everywhere. Birdwatching is free (until you get hardcore about it). And there is so much we can learn from simply sitting and watching birds.

Light-mantled
albatrosses.

Acknowledgements

To the many bird experts, for being so forthcoming and passionate with your avian knowledge. To editor Matt Turner, ornithologist Colin Miskelly and my Penguin Random House crew — Jeremy Sherlock, Sarah Yankelowitz, Laura Sarsfield, Rachel Clark and Cat Taylor — thanks for your amazing input. To Edin, for sharing the journey and bringing the birds to life. And most of all, to my husband Dave, my parents Meg and Bob, and parents-in-law Ferne and Neville. I'm ever grateful for all the support, newborn-holding and toddler-wrangling that made writing this book possible! And to my children — this book is for you.

—Skye Wishart

First, to my parents, who are my steadfast supporters in all things, I can't thank you enough for the life you have given me. This is for you. To Chris Gaskin, for the question *Do you want to photograph New Zealand storm petrels?* and all the adventures that have followed — I've loved every second. You've made so much of this possible. To all those who have ferried me across the waves in search of seabirds and remote islands. To Rodney Russ, Dave Bowen and the Heritage Expeditions crews I've shared subantarctic adventures with. To those who have joined me along the way. To Jeremy, for seeing something worth exploring, and to Skye, for exploring it with me. Thank you all.

—Edin Whitehead

PENGUIN

UK | USA | Canada | Ireland | Australia
India | New Zealand | South Africa | China

Penguin is an imprint of the Penguin Random House group of companies, whose addresses can be found at global.penguinrandomhouse.com.

 Penguin
Random House
New Zealand

First published by
Penguin Random House New Zealand, 2019
This edition published by
Penguin Random House New Zealand, 2023

3 5 7 9 10 8 6 4 2

Text © Skye Wishart, 2019
Photography © Edin Whitehead, 2019

The moral right of the authors has been asserted.

For use of the excerpt from Denis Glover's poem, 'The Magpies' (on page 31), the publisher and authors wish to thank Rupert Glover, of the Denis Glover estate, and the copyright holder, Pia Glover.

Design by Rachel Clark and Cat Taylor
© Penguin Random House New Zealand
Prepress by Image Centre Group
Printed and bound in China by RR Donnelley
Produced using vegetable-based inks

A catalogue record for this book is available from the National Library of New Zealand.

ISBN 978-1-77695-062-1

penguin.co.nz